The Complete Guide to Mercury Toxicity from Dental Fillings

How to find out if your silver dental fillings are poisoning you and what you can do about it.

JOYAL TAYLOR, D.D.S.

Scripps Publishing, San Diego, California

Published by:

Scripps Publishing
9974 Scripps Ranch Boulevard
San Diego, California 92131

Library of Congress Catalog Card Number: 88-90719
ISBN 0-944796-36-2

WHAT OTHERS ARE SAYING
ABOUT THIS BOOK:

Acknowledgements

I would like to take this opportunity to recognize the many who have shared in the creation of this book. Without the able secretarial support of Ann Cummings and editing by Mary Jane Brandes and Lisa Taylor, this manuscript could not have come to completion. Also, photographic assistance was provided by David Collins and cover design by Ed Roxburgh.

I am especially grateful to the many patients who have given me the confidence and allowed me the experiences necessary to write a book on such a controversial issue. A great indebtedness is owed to the researchers and health care providers who have given me the background on this subject matter.

I sincerely thank all these fine people. I know they are proud of the part they have played in the development of the anti-mercury movement in dentistry as well as their contribution to this work.

4-21-89 Quality 14.95

WARNING - DISCLAIMER

The information in this book is intended to complement the advice of your dentist and your physician. The author and publisher assume no responsibility for any adverse outcomes which derive from your use of any material contained within. Any suggestions made in this book should be regarded as general background information to guide you, your dentist, and your physician towards a safe treatment plan. Use of this information without the advice of a skilled health professional is highly discouraged.

WARNING - DISCLAIMER

Contents

Introduction

Toxicity is defined as a pathological condition caused by the action of a poison. The words toxicity, toxic, poison, poisonous and intoxication are used more or less interchangeably throughout this text.

For centuries, mercury has been known to be one of the most poisonous materials available to man. It is easily absorbed into the body directly through the skin with a mere touch. This is why you may recall having been warned to be very careful when disposing of the mercury from a broken thermometer.

*What most people do not know is that the common silver dental filling (also called an **amalgam**) contains 50% mercury, and that this highly toxic metal slowly leaches out of these fillings and gets into the body where it can poison its victims.*

I refer to dental amalgam as silver mercury filling material since this term better describes its true composition.

You may be asking yourself, "How can a known toxic substance, like mercury, be allowed to be placed in people's bodies as in the case of silver dental fillings?" There are many possible answers to this question.

First, the manufacture and sale of silver mercury filling products to dentists is a multi-million dollar business. Manufacturers of these products are not likely to stop producing and selling mercury filling materials unless dentists discontinue using them or the government forbids their sale.

Second, although dentists have several other less detrimental filling materials to choose from, none are as inexpensive and easy to use as silver mercury filling material.

Third, and most significant, is that organized dentistry may not be willing to admit that it has continued to make this tragic mistake for over 160 years. Dentists are still rationalizing their use of mercury based on *unscientific* statements by their own dental "trade" organizations. Here are a few quotes taken from "trade" literature followed by comments regarding scientific research refuting these statements:

> "When mercury is combined with the metals used in dental amalgam, its toxic properties are made harmless."[1] *The scientific literature, on the other hand, indicates that any form of mercury is poisonous.*[2]

> "Although cases of allergy to mercury have been reported in the literature, the prevalence of mercury allergy is estimated to be less than 1%."[3] *A search of the scientific studies, however, shows that as much as 16% of the population may be allergic to mercury.*[4]

"The amount of mercury released is not enough to be harmful to patients unless They are hypersensitive (allergic) to mercury."[5] *However, scientists point out that the levels of mercury vapor from silver fillings approach and, in some cases, exceed the maximum safe allowable limits of the Environmental Protection Agency (EPA).*[6]

As you can see from the above examples, it is not difficult to understand why mercury is used in dentistry.

The toxic effects of silver mercury dental fillings have been discussed and debated in professional circles since mercury's introduction into dentistry in 1819. Several professional books have also been written on this subject. Furthermore, there are more than 1,400 studies found in a number of professional journals concerned with the toxic effects of silver mercury dental fillings.

Not one scientific study in any of those journals proves the safety of silver mercury fillings.

This book has been written in a straightforward, easy-to-understand manner. To facilitate easier reading, the majority of references have been confined to the appendices. Of particular concern to those readers interested in the chemical and biological aspects of mercury intoxication, is Appendix A entitled "Mercury's Effect on Your Body." Also a recommended reading list may be found in the back of this book for those of you who would like to do further research into mercury toxicity. Within the listed books are hundreds of more references concerned with the toxic effects of silver mercury dental fillings.

Whether you decide to have your fillings removed and replaced or simply refuse to allow the placement of any new silver mercury fillings in your mouth in the future, speak out and let your dentist know your opinion about the use of silver mercury fillings. For only you, the general public, can prevent dentistry's continued use of this awful poison, mercury.

REFERENCES

[1] Volume 15, No. 4, *ADA News*, January 2, 1984.

[2] Volume 14, *Environmental Research*, 1977, p. 329-373.

[3] Mitchell, E., *NIDR/ADA Workshop on Biocompatibility of Metals in Dentistry*, July 11-13, 1984.

[4] Nebenfurer, L, et al, "Mercury Allergy in Budapest," *Contact Dermatitis,* 1983, 10 (2): 121-122.

[5] Mitchell, E., Ibid.

[6] Vimy, M., et al, "Serial Measurements of Intra-oral Air Mercury: Estimate of Daily Dose from Dental Amalgam," *Journal of Dental Research*, 1085, 64:1072-1075.

CHAPTER 1

Common Symptoms of Mercury Toxicity

C linical research scientists and dentists alike are continuing to add to the ever increasing list of symptoms of mercury toxicity. At first glance this list seems to include every symptom of unknown origin imaginable; but considering the fact that mercury, like other poisons, can alter the function and processes of every cell of the body, the extent of this list isn't quite as difficult to accept. This book is not implying that mercury is the only cause of these symptoms. It is impractical to assume that symptoms are caused by only one factor. Indeed, symptoms of ill health and chronic diseases, for all practical purposes, are usually multi-faceted.

Before mercury toxicity can be indicated, a dentist or other health care provider must obtain an accurate health history. Occasionally, a mercury toxic patient may have only a few of the symptoms. But usually he will have several. Many patients will check off half or more of the symptoms. These are the patients who must proceed with this treatment slowly. They must not cut any corners with therapy and should be treated only by health care providers who are the most highly qualified in the area of mercury toxicity.

As you read the following list of symptoms of mercury toxicity, you may ask yourself how mercury can possibly be related to some of them. Keep in mind that mercury can inhibit enzymes and hormones, affecting all areas of the body. (See Appendix A).

See how many of these symptoms apply to you. If you answer "yes" to several of them and you have a significant health problem, you may want to consider further testing and evaluation. Also, if you have seen several health care providers for help with these symptoms and have not received satisfactory treatment, you are encouraged to pursue this matter further.

BEHAVIORAL AND PSYCHOLOGICAL SYMPTOMS

Anxiety ✓
Apathy
Confusion ✓
Depression ✓
Emotional instability ✓
Fits of anger ✓
Forgetfulness ✓
Hallucinations
Inability to concentrate ✓
Irritability ✓
Lack of self-control ✓
Loss of self-confidence ✓
Lowered intelligence ✓
Manic-depression
Nervousness ✓
Nightmares ✓
Psychological disturbances ✓
Short attention span ✓
Short-term memory loss ✓
Sleep disturbances ✓
Tension ✓

Trouble making decisions
Unexplained suicidal thoughts
You wish you were dead
Your Doctor said, "It's your nerves."

CARDIOVASCULAR SYSTEM

Abnormal electrocardiogram (EKG)
Anemia
Angina ✔
Arteriosclerosis
Bradycardia
Heart attack
Heart murmur ✔
High blood pressure ✔
Irregular heartbeat ✔
Low blood pressure
Pressure in chest ✔
Tachycardia
Unexplained chest pain

CENTRAL NERVOUS SYSTEM

Chronic headaches ✔
Convulsions
Difficulty in walking ✔
Dim vision ✔

Dizziness ✓
Epilepsy
Facial twitches
Failure of muscle coordination ✓
Hearing difficulty
Insomnia ✓
Loss of ability to perform movements of the hand
Mental disability
Muscle paralysis
Muscle twitches ✓
Multiple Sclerosis
Narrowing of field of vision ✓
Noises or sounds in head ✓
Numbness of arms and legs ✓
Ringing in the ears
Speech disorders ✓
Tingling of fingers, toes, lips, nose ✓
Tremors of hands, feet, lips
Unexplained leg jerks ✓
Voices in head

DIGESTIVE SYSTEM

Colitis ✓
Constipation
Diarrhea ✓
Digestive problems ✓

Diverticulitis ✓
Frequent bloating ✓
Frequent heart burn ✓
Loss of appetite ✓
Stomach cramps ✓
Ulcers

ENDOCRINE SYSTEM

Arthritis ✓
Chronic low body temperature ✓
Cold hands or feet ✓
Decreased sexual activity ✓
Diabetes
Diabetic tendency
Edema ✓
Frequently get up at night to urinate ✓
Increased sweating ✓
Kidney Stones
Leg cramps ✓
Osteoporosis ✓
Pain in joints ✓
Slow healing
Thyroid Dysfunction
Weight loss

ENERGY RELATED SYMPTOMS

Chronic fatigue
Drowsiness
Hypoglycemia
Lack of energy
Lethargy
Muscle weakness
Over-sleeping
Tiredness
Tired when awakened in morning

IMMUNE SYSTEM

Allergies
Asthma
Cancer
Candida Albicans (yeast)
Chronic Fatigue Syndrome
Environmental Illness (E.I.)
Epstein-Barr Virus
Hodgkin's Disease
Immune deficiency diseases
Leukemia
Mononucleosis
Rhinitis

Sinusitis ✓
Susceptibility to flu, colds, etc.
Swollen glands (lymph nodes)

MISCELLANEOUS

Emphysema
Irregular breathing
Kidney damage
Persistent cough

ORAL CAVITY

Bad breath ✓
Bleeding gums
Bone loss around teeth ✓
Burning sensation in mouth
Enlarged salivary glands
Increased flow of saliva ✓
Leukoplakia
Loosening of teeth ✓
Metallic taste in mouth ✓
Mouth ulcers
Periodontal (gum) disease
Purple-black pigments in gums ✓

Stomatitis
Swollen tongue
Tendency towards tartar formation (calculus)
Unexplained loss of teeth
Unexplained sore throat, no infection

SKIN

Acne ✓
Dermatitis ✓
Excessive itching ✓
Rough skin ✓
Skin flushes
Unexplained skin rashes ✓

If you answered "yes" to several of the proceeding symptoms, you may want to examine the role mercury toxicity can play in the human body.

Of course, many health-minded readers may choose to have their silver mercury fillings removed merely because they do not want a potentially hazardous material in their teeth.

Each person must decide if wearing silver mercury fillings is worth the health risk, especially since there are alternative filling materials.

CHAPTER 2

How Mercury Gets out of Your Fillings and into Your Body

T here are four ways mercury can get into your body from silver mercury fillings. Mercury vapor released by the fillings can be absorbed into the *lungs* upon breathing or picked up by nerve endings and blood vessels in the *nose*. Also, the filling material itself can be swallowed into the *stomach* or absorbed through the mouth by *roots*, *bones*, and *gums*. Each of these areas deserves a separate discussion.

LUNGS

As late as 1984 dentists were told that the mercury in silver mercury fillings did not escape. They were led to believe that the normal "evaporation stops as soon as mercury becomes coated with saliva." This hypothesis has been completely disproven, without question.

Mercury is highly volatile, which means it constantly vaporizes (evaporates) from the surface. The amount of mercury vapor from the surface of silver mercury fillings, whether they are coated with saliva or dry, can now easily be determined and measured by using an instrument called a "Mercury Vapor Analyzer" by the Jerome Company. Due to its high cost, only a few dentists have this instrument at this time. But those who do are causing many people to stop and take note. Dentists and patients alike are startled when they see this machine measures mercury as it vaporizes off the surface of silver fillings.

The analyzer takes a small sampling of air from either the mouth or from the breath. The air is analyzed and a digital display measures the amount of mercury in each sample. Persons with no silver mercury fillings usually register no mercury at all; however, people with silver mercury fillings almost always show readings of mercury vapor. If a person with fillings chews something (e.g. gum) for ten minutes, they show levels of mercury vapor 6-15 times higher than before chewing. These "after-chewing" levels will continue for an hour and a half before returning to their former level. "After-chewing" levels can run as high as *one hundred (100)* micrograms or more per cubic meter. This is a justifiable concern since the Environmental Protection Agency's (EPA) maximum safe allowable limit in the *residence* is only *one (1)* microgram per cubic meter. *The Occupational Safety and Health Agency's (OSHA)* maximum safe allowable limit in the *work environment* is only *fifty (50)* micrograms per cubic meter. (More information about maximum safe limits of mercury vapor in Chapter 3.)

Once you see that mercury vaporizes from silver mercury fillings, you can easily understand how breathing introduces it into the lungs. Scientific studies show that 80% of the mercury vapor from silver mercury fillings is absorbed by the lungs.

A related study in Russia was performed on rats that had never previously been exposed to mercury. The rats were placed in an environment with mercury vapor for only 10 minutes. They were immediately sacrificed, and biopsies were performed on various tissues. Extremely high levels of mercury were found in the lung tissue. Levels of mercury almost as high were found in the rats' kidneys, and slightly less were found in the brain and liver. The conclusion was that when rats breathe mercury vapor, their lungs absorb the mercury. The blood picks up the mercury from the lungs and carries it to the kidneys, which try to filter out the poison but absorb large amounts in the process. The brain and liver also absorb considerable mercury from the blood.

Does the same thing happen in humans? We have not had anyone volunteer to sacrifice their lungs, kidneys, brain, and liver for the sake of science; however, there have been several similar studies performed on human cadavers.

Those cadavers with silver mercury fillings had higher levels of mercury in the lungs, kidneys, brain, and liver tissue than did those cadavers without silver mercury fillings.

The facts above should be enough alone to convince most people of the potential danger of silver mercury fillings; but there is more. Read on!

NOSE

As already mentioned, mercury vaporizes from the surface of silver mercury fillings. A high level of this mercury vapor finds its way into the nasal cavity and is absorbed by the nerve endings of small receptors found there. Once inside the nerve cell, it travels up the nerve and into the olfactory bulbs which are extensions of the brain that contain large numbers of nerves which provide us with the sense of smell. From there it has easy access to other parts of the brain, including the hypothalamus and pituitary glands.

Mercury can also be easily absorbed by blood since there are large numbers of arteries and veins in close proximity to the surface of nasal passages. Of particular interest here is the presence of the "Cranio-Vertebral Venous System," a collection of veins surrounding the brain and spinal cord. This system has a peculiar characteristic in that its blood flow can take place freely in every direction. There are no valves to keep the blood flowing in one direction, therefore, the blood *bathes* the brain and spinal cord. The startling thing about this system is that there is no membrane to shield poisons from the brain tissues. Thus, mercury finds a relatively easy route to the brain tissue by way of the nasal cavity.

These two phenomena have been well documented, and there is an entire book written about this subject. Refer to the recommended reading list at the end of this book for *Mercury Poisoning From Dental Amalgam — A Hazard to Human Brain* by Patrick Stortebecker.

STOMACH

If you have had silver mercury fillings for several years, you have probably experienced something similar to the following: You are chewing something (anything from mashed potatoes to steak) when suddenly you hear and feel an uncomfortable "crunch." At first you may think you have broken a tooth, but you don't feel the excruciating pain you might expect. After searching your mouthful of food, you find one or more pieces of a black, hard material. At this point you realize that part of a filling has fractured, and you are relieved that it was not your tooth. Later, you go to a dentist who tells you that fillings must be replaced from time to time because "they wear out and break down eventually." What most people don't stop to think about is the fact that the fillings are "wearing out and breaking down" constantly from the moment they are placed in your teeth. Researchers have determined, on the average, a silver mercury filling "wears out" and needs to be replaced after only 8.5 years. Some people have referred to this fact as "built-in obsolescence" but I prefer to call it *"chronic renewable poisoning."*

"Wearing out" and "breaking down" are less harsh terms than "rusting," "corroding," or "tarnishing," but the latter three terms describe more accurately what is happening. Have you ever had to polish silver because it tarnished? Or found a tin can that rusted after being left out in the rain? Or, have you wondered why new copper pennies are bright and shiny but older ones are dark and dull? The reason these metals have changed is because they have oxidized, or corroded.

What actually happens when a metal corrodes? Elements of dissimilar metals combined with oxygen and acids form totally new compounds called corrosion products. These corrosion products have different physical properties from the original metals and are usually more brittle. (Dozens of corrosion products from silver mercury fillings have been identified, but at this point, very little research has been conducted regarding their toxic properties.)

In the case of silver mercury fillings, mercury is combined with a number of other metals including silver, tin, copper and zinc. Within a few days after the smooth, silver-grey filling is placed, it begins to turn black. And, as the filling is exposed to the acids and temperature changes of the various foods we eat daily, it starts to oxidize even faster. Also, within a few months, its original smooth surface becomes rough. Microscopically small pieces of the filling material begin to fracture away from the edges of the filling. And, as the corrosion continues, larger pieces break away leaving "ditched-out" areas around the edges. Depending on the age of the filling material and its particular composition, these "ditched-out" areas can actually become deep enough to catch food. Finally, an area large enough for you to notice is displaced. But by the time you are aware of the slowly fracturing filling, much harm could have already transpired. You have been swallowing the metals along with your food ever since the fillings were originally placed. And once the mercury, other metals, and corrosion products have entered your stomach, your blood stream has already had ample opportunity to pick them up and transport them to other parts of your body. This is one of several ways your fillings are causing a chronic exposure to at least one known poison, mercury.

ROOTS, BONES, AND GUMS

Researchers who have tested for mercury content in the roots of teeth containing silver mercury fillings have found extremely high levels of this poison, the highest amounts being in teeth which have silver mercury fillings covered with crowns. Since the crown covers the filling, the mercury cannot vaporize, so it is absorbed more rapidly down into the roots. Some clinical researchers estimate that 80% of the crowns on back teeth have silver mercury fillings under them. These can be porcelain, gold or nickel crowns. Roots of teeth containing silver mercury fillings covered with gold crowns show the highest mercury levels of all.

High levels of mercury are also found in bones surrounding roots of teeth containing silver mercury fillings. Even the gum tissue around teeth with these fillings shows consistently high levels of mercury. It is a well known fact that amalgams are one of the contributing factors in gum disease (pyorrhea). Once mercury is absorbed by the rich supply of blood vessels and capillaries found in the gums, roots and bone surrounding the teeth, the blood stream can carry it to all parts of the body.

As you have read, mercury comes out of the filling and can enter the body through several routes. No one knows how much mercury is absorbed by each route. However, the quickest and most efficient route for a body to absorb almost anything is through the nose (e.g. snorting cocaine) and under the tongue (e.g. sublingual absorption of nitroglycerine). Mercury from silver mercury fillings has easy access to both routes. The important thing to understand is that separately, each absorption route of mercury from silver fillings poses a potential threat, but collectively they may represent a clear and present danger to your health.

CHAPTER 3

How You Can Be Tested for Mercury Toxicity

Until recently, the only methods used to diagnose mercury poisoning were the questioning of patients about their symptoms and the observation of certain physical and emotional signs. These two methods are still quite necessary; but now there are additional tests and screening procedures that serve as further indicators.

No single test can be used to make a diagnosis of mercury toxicity.

However, collectively several tests can reveal indicators that may support toxicity. Some of these tests are repeated periodically after your silver mercury fillings have been removed in order for your health care provider to monitor your progress.

LABORATORY TESTS

Various laboratory tests can be performed in conjunction with mercury toxicity (Fig. 3-1). These tests serve many purposes:

1. to indicate the effects of mercury,

2. to monitor a patient's progress,

Fig. 3-1. Laboratory Tests: Collectively, blood (top), urine (center), and hair (bottom) testing can reveal many indicators of mercury toxicity as well as imbalances of body chemistry.

3. to determine what dental materials the patient is best able to tolerate,

4. to evaluate overall health status, and indicate nutritional modifications needed,

5. to recommend vitamins and minerals for mobilizing and excreting the mercury, and

6. to help balance the body chemistry.

The National Institute of Occupational Safety and Health has stated:

"Although mercury levels in the blood, hair and urine do not show direct correlations to the appearance of symptoms, these laboratory tests can be invaluable in assessing indirect affects of mercury on the body as well as indicators of the body's excretory efficiency. Mercury's potential to effect the immune system, enzymes and thyroid function are examples."

Blood Tests

SMA and CBC blood tests help to determine how mercury may be affecting various parts of the body. For example, mercury can contribute to fatigue by attaching itself to red blood cells and compromising their ability to carry oxygen. It is also believed that mercury can raise glucose and lower cholesterol. Additionally, there are many other effects suggestive of mercury that can be seen in SMA and CBC blood tests, but they are too comprehensive to discuss in a book of this type.

Monitoring the level of mercury itself in the blood is not routinely done because the blood can clear itself of the majority of mercury from new exposures in a relatively short amount of time. However, tests have shown that the average level of mercury in the blood of persons having silver mercury fillings is more than twice the level of persons having no silver mercury fillings.

Urinalysis

Urine tests specific for mercury are helpful in determining if mercury is being adequately excreted by the body. Since everyone is exposed to some mercury from air, food, and water daily, a certain amount of mercury excretion is expected. A high level of mercury in the urine suggests that the body's mechanisms of excretion are operating efficiently. This may also indicate, however, that the body has had a heavy exposure to the metal. Surprisingly, a low urine mercury level may indicate the most problems. If there has been a fair amount of mercury exposure and the urine mercury level is too low, the excretion mechanisms may not be functioning properly, which would suggest a condition whereby the patient is *retaining mercury*. Steps may then be taken to mobilize the mercury which is being retained. Some health care providers may assume that the lower the urine mercury levels, the better. As we have just discussed, the proper interpretation of these tests is essential. Also, since mercury is absorbed by glass, special precautions must be taken with the handling, storage and transportation of urine samples for determining accurate levels.

Hair Analysis

Hair analysis is used as an indicator for lead, cadmium, arsenic, and mercury poisoning, as well as other heavy metals. Only a small sample of hair is needed for the test. The sample needs to be taken from the nape of the neck on the back of the head. The first two or three inches closest to the scalp gives the most recent hair growth. Follow-up hair analyses taken every 3 to 6 months allow the health care provider to monitor the patient's progress.

Hair analysis can also be used in conjunction with blood and urine tests to monitor the effects of mercury on the patient's mineral balance. In addition, proper supplementation programs can be enhanced with the help of hair analysis.

T-Cell Analysis

T-Cells are the white blood cells associated with antibody regulation and the immune system. Mercury has a negative effect on the immune system, in general, and silver mercury fillings have been shown to reduce the T-cell percentages. Although very few health care providers order the T-Cell Analysis, I recommend this specialized blood test for patients before and after the removal of silver mercury fillings. An improvement toward a more healthy immune system is one of the first responses noticed when treating mer-cury toxicity. The T-Cell analysis is, therefore, one way to monitor a patient's progress.

Biocompatibility Testing

As this book is being written, there is a new blood test being developed to check for compatibility of various dental materials for each individual. I have had the opportunity to personally observe this technique. I feel that there is great promise in biocompatibility testing and I believe that this test has the potential of revolutionizing dentistry.

SKIN PATCH TEST FOR
MERCURY SENSITIVITY

Skin patch tests are routinely used by allergists, environmental ecologists, and dermatologists to help determine allergies and sensitivities to foods and chemicals. A small amount of the substance being tested is placed on a bandage or gauze and taped to the skin, usually on the arm. The patient wears the patch for a designated period of time, and upon removal of the patch, the doctor checks the skin for allergic responses (reddening, itching, etc.).

When testing for mercury sensitivity, the above procedure becomes more complicated. Blood pressure, pulse rate, and body temperature are recorded before administering the patch and during the first hour of the test. The patient remains seated and relatively inactive during this period and isn't allowed to eat, drink, smoke, take medication, or engage in any activity that may alter the physiological systems that are being monitored. A local skin reaction is observed in about one-third of the population, but more importantly, are the generalized body reactions that occur. Significant changes in the blood pressure, pulse rate, and body temperature during this one hour "quiet" period, are indications that mercury may be affecting the body on an overall or systemic basis.

There can also be an exaggeration of the patient's symptoms during this testing period. For example, the patient may experience muscle twitches, headache, emotional outbursts of anger or crying, depression, pain in the stomach or joints, etc. Because of the likelihood of an increase in symptoms, a patient should not be left alone during this testing period.

The patch is removed immediately whenever symptom indicators become evident. The area is then washed with soap and water and the patient is given 1 1/2 tsp. of sodium ascorbate powder dissolved in water to counteract the reaction.

If there are no reactions during the one hour "quiet" time, the patient continues wearing the patch at home for a 24-hour maximum period of time. They are instructed to watch for exaggerated symptoms during this time. Another person should constantly be with the patient during this time period, and sodium ascorbate powder should be available, if needed.

Those patients who react during the first hour are most sensitive and are primary candidates for silver mercury filling removal.

Skin patch tests with mercury are not always performed due to the inherent dangers of the test, such as:

1. the unwanted additional exposure of mercury to both the patient and the doctor,
2. the chance of sensitizing a patient to mercury,
3. the stress of patients having to deal with exaggerated symptoms, and
4. the unwillingness of "mercury-free" dentists to stock this poison in their office.

The test is usually only offered to those patients who specifically ask for such a procedure or for those who can't decide whether they should have their silver mercury fillings removed. Patch testing can only help in determining *allergies* (or hypersensitivities) to mercury.

Since allergic responses and toxic (poisonous) potentials are totally different problems, patch testing with mercury has limited value.

ELECTRICAL MEASURING DEVICES

Dentists involved with mercury toxicity usually have access to one or more electrical measuring devices. These instruments measure the voltage and amperage of the electricity generated by the metals in your mouth. This information allows the dentist to remove the silver mercury fillings in a preferred sequence. It is not well understood why some fillings are said to have negative currents and some have positive ones. However, one well known clinical research dentist has discovered that removing the highest negatively charged fillings *first* produces the most satisfactory results. In fact, removing the positively charged fillings *first* is said to have caused an increase in negative health symptoms in some cases.

MERCURY VAPOR
ANALYZER SCREENING

The *Jerome Mercury Vapor Analyzer* is an instrument designed to measure the amount of mercury vapor in the air (Fig. 3-2). It is important that you understand how it works. This machine draws in a time-measured amount of air through a small tube. The air passes over a gold strip inside the box. Any mercury in the sample of air is attracted to the gold and attaches itself like a nail to a magnet. The electrical capacity of the gold is changed depending on the amount of mercury present and the computer within the box can displays a reading of how much mercury is present in the sample of air.

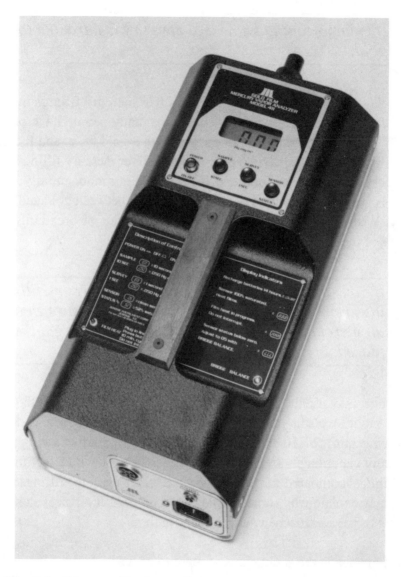

Fig. 3-2. Mercury Vapor Analyser: Used to detect mercury vapor from silver fillings. This instrument is also used by The Occupational Safety and Health Agency as well as other organizations. (Photograph courtesy of The Arizona Instrument Corp., Jerome Division).

> *It is important to know that mercury is attracted to gold.*

Let's say that you have both gold and silver mercury fillings in your mouth. This situation is not unusual since many people have one or more gold crowns or porcelain crowns with a gold base, and the majority of people in the civilized world have silver mercury fillings. The mercury vaporizes (evaporates) from the filling more rapidly when gold is near because there is a natural attraction between gold and mercury. With this situation, there is a greater chance of breathing high levels of mercury vapor.

> *If you are going to have your silver mercury fillings removed and replaced with gold, be sure that all mercury fillings are removed before any gold is placed.*

A dentist may use the Jerome Mercury Vapor Analyzer to test the mercury vapor levels in your mouth. I suggest a test both before and after chewing gum for 10 minutes. On the average, after chewing gum, mercury vapor levels are 54 times greater in persons with silver mercury fillings[1] compared to persons without silver mercury fillings. This fact may explain why some people with mercury fillings feel sick after eating or get a headache when chewing gum.

To demonstrate the seriousness of mercury contamination, I would like to discuss tests performed in other contaminated environments. Various agencies use the Jerome Mercury Vapor Analyzer to test for mercury contamination. For instance, the *Occupational Safety and Health*

Agency (OSHA) is allowed to test the mercury vapor levels of any business that uses mercury, for example, a thermometer manufacturing company. Periodically, OSHA will test the mercury vapor levels of air around the employees who are actually handling the mercury. If the analyzer shows the levels to be above a certain amount, that company will be closed until safe levels can be established. In addition, they are fined $10,000 and all employees are monitored for 5 years for mercury levels in the blood and urine.

Presently, OSHA uses 50 micrograms per cubic meter as their maximum allowable levels of mercury vapor in the work area.[2] This number usually has no significance to most people until they see the amount of mercury vapor in their own mouth, especially after chewing gum. For people with several silver mercury fillings, the average levels are 50 to 150 micrograms per cubic meter.[3] This is one to three times the maximum allowable levels by OSHA. Levels as high as 400 micrograms per cubic meter have been reported while testing the vapor around one single, silver mercury filling.[4] By all rights, many people's mouths should be shut down, and those people should be fined $10,000 until the mercury is their mouths is cleaned up. Also, OSHA is presently entertaining the idea of lowering its maximum allowable level from 50 to 20 micrograms per cubic meter.

The Environmental Protection Agency (EPA)[5] and NASA[6] use 1 microgram per cubic meter as their maximum allowable level of mercury vapor.

Various European countries,[7] the Soviet Union,[8] and the U.S. Navy[9] have 10 micrograms per cubic meter as their maximum allowable levels of mercury vapor. The National Institute of Occupational Safety and Health (NIOSH) uses 20 mcg/m^3 as their maximum safe allowable limit and the *World Health Organization (WHO)* uses 25 mcg/m.[10,11] One of the few places where mercury is used but not periodically tested is in dentistry. Neither *OSHA* nor the *EPA* monitors the mercury levels in dental offices and most dentists have never heard of the Jerome Mercury Vapor Analyzer. I have one of these instruments and have tested many dental offices, finding over 50% of the dental offices tested with mercury vapor levels to be over the maximum allowed by *OSHA*, and approximately 98% over the *EPA*'s maximum level. This may be part of the reason why so many people feel ill just walking into a dental office.

If you do not have symptoms and are considering having your silver mercury fillings removed for preventive reasons only, you may need very few of the previously mentioned tests.

However, for those of your who have severe problems or multiple symptoms, full testing is recommended. Everything possible should be done to increase your chances for recovery.

REFERENCES

[1] Vimy, M., et al, "Intra-oral Air Mercury Released from Dental Amalgam," *Journal of Dental Research*, August 1985, 1069-1071.

[2] O.S.H.A., "Job Health Hazard Series: "Mercury," OSHA 2234, August 1975.

[3] Utt, H., "Mercury Breath . . . How Much is Too Much?" *CDA Journal*, February 1985, p. 42.

[4] Ibid.

[5] EPA, "Mercury Health Effects Update: Final Report," United States Environmental Protection Agency, Office of Health and Environmental Assessment, August 1984, EPA-600/8-84-019F.

[6] Jerome Instrument Corporation, op cit.

[7] Stortebecker, P., *Mercury Poisoning From Dental Amalgam — A Hazard to Human Brain*, Stockholm, Sweden: Storebecker Foundation for Research, 1985.

[8] Friberg, L., et al, Mercury in the Environment, Cleveland, Ohio: CRC Press, 1972, p. 110.

[9] Jerome Instrument Corporation, op cit.

[10] W.H.O., "Recommended Health — Based Limits in Occupational Exposure to Heavy Metals," Report of a World Health Organization Study Group, Technical Report, Series 647, W.H.O., Geneva, 1980.

[11] N.I.O.S.H., "Recommended Standard for Occupational Exposure to Inorganic Mercury," published by NTTS, No. PB-222,223, 1973.

CHAPTER 4

How To Select the Best Filling Materials for You

T oday there are many dental materials that can be used as alternatives to silver mercury fillings: porcelain, gold, composite (plastics), and other metals. This chapter is devoted to a discussion of both the advantages and disadvantages of each material. First, however, one fact needs to be understood.

There is no perfect man-made dental material and there never will be.

Man, in all his wisdom, will never be able to adequately replace his natural God-given tooth structure.

Now let's take a look at the man-made dental materials available, other than the silver mercury filling.

PORCELAIN

Porcelain is, by far, the best material available today. It is not perfect; but it has the most advantages and the fewest disadvantages of any other material, and it has been used for many years with varying degrees of success. Most of the weakness with porcelain in the past (hardness, brittleness, etc.) have been greatly improved. It can now be used in small or large fillings, inlays and onlays, crowns, and even small bridges. Porcelain is tinted the color of a tooth so you cannot usually tell where it is. Even dentists are sometimes fooled during a visual examination of a patient with porcelain restorations. Porcelain does not stain, and it is generally considered a "permanent" restoration since it does not corrode or wear down. Porcelain usually does not register electrical conduction. This is a very desirable characteristic since it has been concluded that electricity generated by dental restorations is an irritant that may cause oral diseases, including malignancies.

There are presently several manufacturers of dental porcelain, and each has its own "secret" formula. Porcelain may contain metal oxides (particularly aluminum oxide); but some brands contain less than others.

Porcelain fillings, crowns, etc., are usually placed in two appointments, since an impression of the tooth must be made and sent to a dental lab where the porcelain is baked in a special oven. With the advent of computerized laser technology, however, the second appointment and the extra time necessary for laboratory fabrication may be eliminated in the near future. France has a "box" that is reported to take a 3-dimensional picture of the prepared tooth while laser-cutting a filling or crown out of a block of porcelain. This procedure is done beside the dental chair within minutes while you wait. Consequently, future treatment may become much faster and your investment considerably reduced. Until the use of this advanced technology becomes widespread, however, porcelain will be more expensive than some of the other materials we will discuss. Keep in mind, you get what you pay for. And besides, if cost were the only criteria for choosing a dental material, you would be happy with silver mercury fillings and not be reading this book now.

Many dentists will refer to a crown as porcelain when, in reality, the crown has a metal base with a covering of porcelain. This combination defeats many of the advantages of porcelain; so be certain the dentist understands you want *solid* porcelain. Also, some dentists refer to composites as porcelain. Composites are really plastics with porcelain powder or other components sprinkled throughout. This mixture is not the same as "porcelain."

There are some instances where solid porcelain cannot be used; such as in a case where several teeth are missing and a long "bridge" is needed. The porcelain might have to be reinforced with an underlying metal for support. If this instance occurs, I suggest choosing the highest quality metal alloys which are discussed next under the heading of *gold*.

I recommend porcelain as an excellent choice to consider first for most dental restorations.

GOLD

For centuries gold has been used as a dental material; however, not all dental gold is the same.

When pure gold is used (which is rare) it is called "*gold foil*." Foil consists of extremely thin pieces of gold that are pounded into a tooth cavity. Gold foil can be used on small cavities on non-chewing areas of teeth, but is too soft to be used as a general all-purpose tooth filling material. Also, dentists who are experienced in its use are not common. Gold foil dentistry, was for many years, considered "the state of the art." However, today it is, for all practical purposes, a "lost art".

That is unfortunate, indeed, because gold foil appears to be biologically compatible for many people. These fillings are usually the most expensive dental restorations due to the amount of time involved, the cost of materials, and the highly specialized skills necessary. If you happen to know of a dentist who routinely works with gold foil, you have probably found the rare individual who is a "true" artist. On the other hand, it is unlikely that all of the fillings of any one patient are such that gold foil could be used anyway.

When gold is used in dentistry it is usually combined with one or more metals, such as platinum, palladium, copper, zinc, and silver. This may be where patient's sensitivities enter in. They may believe themselves to be allergic to gold, but it may be some of the other alloys used with it. Sometimes there are small amounts of unusual sounding metals, such as iridium, indium, and gallium. These metals are combined with gold to enhance certain physical characteristics. For example, platinum and palladium are added to gold to increase the hardness. Also, lower quality metals may be added to gold to reduce the cost.

When two or more metals are combined, the mixture is called an *alloy*. Gold fillings, gold crowns, gold bridges, gold partials, and even gold jewelry are really alloys — gold being only one of the ingredients. Sometimes the amount of gold in an alloy is relatively small. Some so-called "golds" have as little as 10% gold. Other alloys of gold contain 80% to 90% gold. The higher the gold percentage of the alloy, the greater the chance of it being accepted by your body. Also, the fewer metals that are in an alloy, the better. When situations arise that require the use of metals for added stability, using alloys with the very highest gold content are suggested.

For most individuals, I do not recommend the use of gold alloys. One reason for this is that when several metals are combined, a complex electrical system begins operating.

No one fully understands the side effects of these "battery cells" to the nervous system and the human body in general. Also, the gold content may be acceptable for most people, however, the other metals may have a negative effect. Science and medicine have only scratched the surface of the toxic effects of well-known metals like mercury, nickel, lead and cadmium within the human body. And, relatively little research has been directed towards iridium, palladium, and gallium. Gold fillings have long been considered the favored dental material; however, more and more dentists and patients alike are beginning to question their biological compatibility.

COMPOSITE (PLASTICS)

A composite is a tooth colored filling which is much less expensive when compared to porcelain or gold. It is quickly replacing silver mercury amalgam as the most used dental material.

Composites are composed basically of plastics, but are probably called "composite" because the public would rather buy something with a more sophisticated name than "plastic."

Regardless of what anyone tries to tell you, a plastic by any other name is still a plastic.

Composites contain all kinds of "toxic sounding" chemicals. Some of the ingredients are petroleum based hydrocarbons. There is bis-phenol glycidyl methacrylate and/or urethane dimethacrylate. There is an alcoholic solution of tertiary amine, ethanol, and solfinic acid salts sometimes present. These compounds are sometimes mixed with things like benzoyl peroxide or benzoine ethyl ether which then cause a chemical reaction, hardening the material. Understand that these chemical reactions usually take place after the materials are placed in your teeth. Powdered materials are mixed into the plastic compound to give the fillings color and strength. Another characteristic that is sometimes necessary is radiopacity. This means that compounds like metal oxides are added to the materials so that the dentist can see the fillings on x-rays. Some of these so-called "fillers" include such harsh sounding things as lithium aluminum silicate, barium aluminum silicate, barium fluoride, quartz silicate glass, and oyster shell. There is even one composite that contains mercury, believe it or not!

With all the concerns surrounding mercury, there isn't the time, money or desire to research the possible side-effects of placing composites into people's teeth.

I predict that after the dust clears on the mercury controversy, equal or greater concern will be directed at the composite.

Composites are also notorious for creating tooth sensitivities and, occasionally, even the death of a tooth. There are many reasons these problems may occur. Many people may be allergic to one or more of the chemicals present in composite or sensitive to the chemical reactions that take place when the material hardens in the tooth. Shrinkage of the material as it hardens may cause internal pressures within the tooth. In deep fillings, the nerve of the tooth is in close contact with the chemicals of composites, and chemical "burns" may result. Air pockets within the plastic may build up pressure when heated by hot foods a patient eats. Finally, the pH (or acidity) of composites may be too extreme for some people's teeth. These are a number of the reasons given by researchers as the causes of tooth sensitivities from composites.

After preparing a tooth for a composite but immediately before placing the material, the dentist has to perform a potentially dangerous procedure. He places a strong solution of phosphoric acid on the tooth for 30 to 60 seconds. This is called "etching," a term for "chemically burning and dissolving" the surface of the tooth. Microscopically, the enamel is cleaned of organic materials and this leaves behind hard mineral constituents of tooth structure with a rough surface. The thinnest component of composite flows into these microscopically rough areas and cements or glues the material to the tooth. The biggest problem with this so-called "acid etching" procedure is that the phosphoric acid may soak into the tooth and irritate the nerve (pulp).

If after placing a composite sensitivities occur, the dentist may try to re-fill the tooth, with a different composite material. If this is not effective the nerve within the tooth may have to be removed. The hollow space left behind is then sealed with a variety of potentially poisonous materials. This is what dentists refer to as a root canal. After treatment, the tooth generally requires a crown due to the large amount of tooth that had to be removed in the "root canal" preparation.

Physically and mechanically the composites present difficult and sometimes compromising situations. The material is extremely technique sensitive and can be frustrating for the dentist to work with as well as time consuming for the patient. For example, a medium size cavity that is filled with silver mercury filling material may require 15 or 20 minutes of the dentist's time. The same cavity, if filled with composite, may require 40 to 60 minutes.

Shaping the hardened composite is time consuming because it has to be done with a drill after the filling is completely set. Silver mercury fillings were easy, in comparison, because they could be shaped with special handcarving tools while the filling was hardening. It is impractical for a dentist to do this with composite material.

When composites first became popular, dentists were charging the same fees for them as they had been charging for silver mercury fillings. The dentists were so happy to be able to offer tooth-colored fillings as an alternative to the ugly black silver mercury fillings that they ignored the extra time, skills, and expense required. Today, it is common for a dentist to charge slightly more for composites than he does for silver mercury fillings.

If you do decide to have composites placed in your mouth, remember this: The larger and deeper the filling, the greater the chances of tooth sensitivity problems.

CEMENTS, BASES, LINERS, AND BUILD-UPS

These materials are the weakest link in the search for biologically compatible dentistry. Whether a patient chooses to have porcelain, gold, or composite, the dentist must use something under these materials to protect the tooth nerve and to cement the fillings to the teeth.

Cements

Cements are used under porcelain, gold, and composite fillings as well as crowns and bridges. They hold the dental material to the tooth. Without cements, the restorations in your mouth would simply fall out. There are many kinds of cements to choose from and each dentist has his favorites. One popular cement called zinc phosphate is very acidic and can cause sensitivity and even death of a tooth.

Liners

To reduce the chances of tooth pain from many cements, a protective "liner" is first placed in the tooth. There are many products used as liners, and their chemical make-up is sometimes a "trade secret." One brand smells like airplane glue, but as far as I know, no one has studied its possible side effects. Another often-used liner is basically calcium hydroxide, but it, too, has other "trade secret" ingredients.

Bases

Bases are materials used to insulate the tooth nerve against temperature changes when there is a filling that is large or close to the nerve. Because they can cause irritation and sensitivity by themselves, bases are usually not placed in direct contact with the deepest part of a prepared tooth cavity but directly over a liner.

Build-ups

Build-ups are materials used to partially re-build badly broken down teeth before placing crowns. Most teeth that have crowns on them also have a build-up material underneath. Unfortunately, most dentists have been using silver mercury filling material as a build-up material. Approximately 80% of the crowns that I remove have silver mercury amalgam underneath. This is an extremely detrimental situation because the mercury vapor is then forced down into the tooth. There are terrifically high levels of mercury in the roots and surrounding bone and gums of teeth with silver mercury filling material underneath a gold crown.

*   *   *   *   *

Notice that there may be four different materials under fillings or crowns (liners, bases, build-ups, and cements), each with its own chemical make-up and potential for allergy, sensitivity, and toxicity.

There may be a partial solution to the dilemma of using several materials under restorations. I recommend using a dental cement called glass ionomer and nothing else. This material has been available from a number of manufacturers for a few years and offers several advantages. It can be mixed thinly for a liner and seems to be more compatible than many other dental materials that are placed directly in contact with the deepest part of a prepared tooth cavity. It can be mixed thicker for a base or a build-up since it has some insulating abilities and considerable strength. Glass ionomer was developed as a cement and actually bonds to tooth structure. It can be used under porcelain, gold, or composite (plastic) fillings as well as under crowns and bridges. It is basically composed of silicate (glass), but it may also contain some less desirable chemicals.

Glass ionomer is not a perfect dental material by any stretch of the imagination, but at least it allows the elimination of three other potentially harmful materials.

Patients and dentists alike may forget that the materials placed under the fillings and crowns come in direct contact with the teeth, and therefore, they are just as important as the material chosen to restore the tooth.

OTHER METALS

During the late 1970's when gold prices skyrocketed, dentists began using other metals for crowns and bridges as a means of reducing costs. These other metal alloys are commonly called "non-precious" metals, which means they will tarnish, corrode and oxidize. Today, even with gold prices down, non-precious metals constitute eighty to ninety percent of all new crowns and bridges.

Non-precious alloys used in dentistry are composed of combinations of metals like nickel, beryllium, antimony, iron, titanium, vanadium, chromium, and cobalt. The greatest percentage of these alloys is composed of nickel.

Of all the metals, nickel has been shown to be the most allergenic and has even been used by scientists to induce cancer in laboratory animals.

It has been known for many years to be a highly toxic metal. Nickel may be as toxic if not ***more so*** than mercury.

Some of the other components of non-precious alloys may also be detrimental to a person's health. Beryllium is a known poison. And, who knows what effects metals like titanium and vanadium may have on the human body?

I suggest that the use of non-precious metals in dentistry be discontinued. Also, those patients seeking optimum health may want to consider the removal of these metals from their mouth and the subsequent replacement with porcelain and/or high gold content alloys.

THE LOWEST COST
ALTERNATIVE MATERIAL

Some patients are on a low, fixed budget and cannot afford to have their silver mercury fillings replaced with porcelain, gold or even composite (plastic). A temporary solution could be what I call "glass ionomer temporary fillings."

After removing the silver mercury fillings, the dentist can simply fill up the cavity left in your teeth with this material instead of only placing a thin coat into the deepest part of the prepared tooth (which is done when glass ionomer is used as a liner under restorations).

The advantages of doing this type of temporary filling are that the procedure is quick and relatively inexpensive. Also, the body now has a chance to respond to the removal of mercury from the tooth. Another advantage is that later, when a more lasting filling is desired or necessary, the dentist may have to remove only a portion of the glass ionomer material, since he will need a thin coat of it in the deepest part of the cavity anyway.

The disadvantages are numerous and should be thoroughly understood before this procedure is considered. These temporary fillings cannot be expected to look, feel, or function like porcelain, gold or composite. They are too white, rough, and wear down quickly. Food may become wedged around them and the bite may not feel quite right. Also, they break and chip relatively easily. In the long run, the cost will be greater because these temporary fillings will have to be replaced within a few weeks or months. Glass ionomer is not strong enough to be used when a crown is removed, or even to replace a very large silver mercury filling.

This type of low-cost dentistry is considered very poor quality by any dentist's standard.

Many dentists will ask you to sign a release form if you have this type of filling placed because patients usually expect more from these temporaries than they can offer. Many dentists refuse to offer this type of service because of the many compromises and inconveniences involved.

Ultimately, you, the patient, will have to decide if the possible health improvements outweigh the many disadvantages of glass ionomer temporary fillings when cost is the overriding factor.

CHAPTER 5

Nutrition, Supplementation, and Lifestyle Factors

(The Critical Links to Successful Therapy)

What, you may be asking, does nutrition, supplementation, and lifestyle factors have to do with mercury toxicity? The answer is: *practically everything*, especially when a chronic exposure to mercury (as from your silver mercury fillings) is being addressed.

Many effects and symptoms of mercury toxicity (see Chapter 1 and Appendix A) cross correlate with effects and symptoms of poor dietary habits and lifestyles. For example, your main symptom may be that your immune system is depressed, making you susceptible to colds, flu and infectious diseases, in general. As previously discussed, mercury may be a contributing factor to immune depression, but so can a lack of sufficient Vitamin C. You may not be eating enough fruits and vegetables that are high in Vitamin C, and if you smoke, your body needs an extra 25 mg. of Vitamin C for each cigarette. Mercury may have a detrimental effect on the endocrine glands, for instance; but so does sugar, alcohol, and caffeine. There are many other similar examples that will be mentioned later. But my main point is: if you want results, help yourself by adopting a healthy lifestyle and good eating habits.

In other words, if you expect the removal of your silver mercury fillings by itself to cure all that ails you, there is a great likelihood that you will be very disappointed!

Health is a condition to be strived for and maintained on a constant basis. Nutrition and lifestyle habits along with heredity and genetic factors are extremely important elements that contribute to good health. Neglecting the role these factors play and only removing your fillings in hopes of improvement is similar to taking a "shot in the dark." The chances of hitting your target may not be worth the risks involved.

Ideally, the removal of your fillings has the greatest chance of helping you if you are already eating healthful foods and have eliminated lifestyle habits that are potentially dangerous. Even those who have had healthy lifestyles for considerable periods of time may need personal counseling because there are many special considerations for mercury detoxification. One example is that during the initial phase of therapy, it is recommended that no fish or seafood be consumed. This is because they can absorb high levels of mercury and convert it into methylmercury (the most toxic form of mercury).

This chapter is designed to give only basic information on the effects of diet and lifestyle as they relate to mercury toxicity. Each person's body chemistry should be evaluated on an individual basis by a qualified health care provider (Fig. 5-1). Only then can a diet and supplement regimen be properly recommended. I suggest this along with removal of silver mercury fillings for optimum results. Every person is different, and nutritional as well as dental recommendations should be tailored to the individual.

If you would like to read an excellent book on good, sound basic nutrition, *The Lazy Person's Guide to Better Nutrition*, by Gordon Tessler, PhD., is highly recommended. This and other recommended nutrition and lifestyle books are listed in the back of this book. Any one of these books will give you a good start towards optimum health.

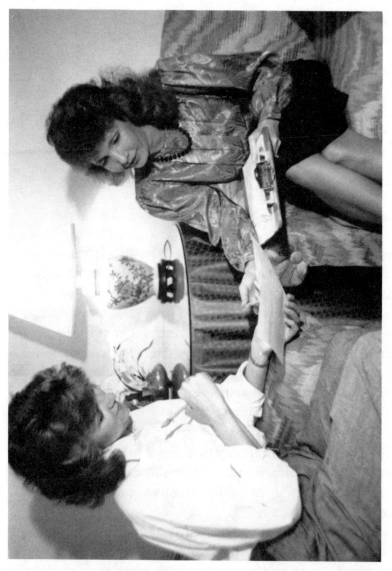

Fig. 5-1. Body Chemistry Counseling: Nutritional, supplemental, and life style factors relative to the patient's body chemistry are addressed during the critically important counseling sessions.

Later in this chapter, certain nutritional and dietary suggestions that particularly apply during mercury detoxification are made.

GENERAL DIETARY AND LIFESTYLE RECOMMENDATIONS

The following are basic factors that can effect a patient's ability to achieve optimum health.

SUGAR

When the average person thinks of sugar, he usually visualizes *white table sugar*. This is a simple sugar that, more correctly, is called sucrose. Other simple sugars are corn syrup, corn sweetener, lactose, fructose, dextrose, brown sugar, turbinado sugar, maple syrup, molasses, and honey. Simple sugars are absorbed very quickly by the blood stream and give a temporary, quick energy lift immediately after they are consumed. Unfortunately, an energy "crash" results approximately 20 minutes later because when too much sugar enters the blood too quickly, the oxygen transported by the blood slows down. Thus, simple sugars are said to rob your cells of needed oxygen. Mercury may have a similar effect because it can inhibit the ability of red blood cells to carry oxygen to the cells (see Appendix A).

Sugars require various vitamins and minerals for their digestion. When the above simple sugars are consumed, various nutrients must be "borrowed" from other parts of the body to aid in the digestion. Simple sugars, therefore, contribute to the depletion of calcium, magnesium, manganese, chromium, and zinc, as well as the B-complex vitamins. The nervous system may be compromised from a depletion of B complex vitamins resulting in irritability and other central nervous system (CNS) symptoms. Also, an imbalance of calcium and magnesium can lead to tooth decay and osteoporosis. A lack of manganese hinders the operation of enzyme systems, and a shortage of zinc results in hormone imbalance. The pancreas is also substantially affected by sugar consumption. If simple sugars are consumed over long periods of time, the pancreas is forced to overproduce insulin in order to control sugar levels in the blood. If overworked, the insulin production mechanism may become damaged, resulting in diabetes. Consumption of simple sugars also raises glucose, cholesterol, and triglyceride levels in the blood.

The average American eats more than 120 pounds of refined, white sugar per year. Some food labels list several sugars in their ingredients, which may collectively comprise the majority of the product. Even some commercial salt contains sugar. "Junk" foods such as soft drinks, candy bars, cookies, cakes, pies, and pastries contain large amounts of simple sugars. The average soda contributes 7 to 8 tsp. of sugar to a person's diet. You probably cannot imagine putting 7 or more tsp. of sugar into a glass of water and drinking it, but millions of people are doing it daily in the form of soft drinks and many other assorted so-called food substances.

Unacceptable sugar substitutes are the artificial sweeteners such as sorbitol, manitol, xylitol, and sodium saccharine. These are chemicals, and they can inhibit cellular efficiency while accumulating in the cells.

Complex sugars, such as those found in whole grains like brown rice, oats, rye, barley, millet, etc., and all types of fruits, are recommended for a healthy diet because they are digested slowly and cause no drastic elevation in blood sugar levels. These complex sugars also contain efficient vitamins and minerals to assist the body in their digestion and do not deplete the body of nutrients like simple sugars do.

The important thing to remember here is that health minded people eat complex sugars and stay away from products made from simple sugars.

CAFFEINE

Caffeine is not only found in coffee, but also in tea, chocolate, aspirin, and many over-the-counter drugs. It contributes to some of the same imbalances in your body chemistry that are caused by simple sugars, namely, an increase in glucose, cholesterol, and triglycerides, and a depletion of magnesium and zinc. Caffeine also raises uric acid, interferes with protein metabolism, and upsets endocrine balance. Note: In addition, coffee contaminates the body with the toxic heavy metal cadmium.

Health minded people do not consume products containing caffeine.

ALCOHOL

Alcohol has most of the same negative effects on your body's chemistry as sugar and caffeine. It raises blood levels of glucose, cholesterol, triglycerides, alkaline phosphatase and LDH. It depletes magnesium, zinc, manganese, potassium, and folic acid and interferes in protein metabolism and endocrine balance. Alcoholic beverages can also contain high levels of sugar and chemical preservatives.

Health minded people refrain from alcohol consumption.

SMOKING

Smoking is an unhealthy habit that raises cholesterol and triglyceride levels in the blood. It interferes with cell membrane metabolism, thus hampering the excretion of mercury. Smoking also contributes to lead and nickel contamination, which can cause symptoms similar to mercury poisoning.

* * * * *

The regular consumption of sugar, alcohol and caffeine along with habitual smoking can all be considered *drug addictions*. These products cause imbalances in your body's chemistry, and not only hamper the excretion of mercury, but can cause signs and symptoms that mimic mercury toxicity.

PHYSICAL ACTIVITY AND REST

Health minded people know the importance of regular physical activity and sufficient rest. However, you may be too sick to exercise and/or your illness may prevent you from getting a good night's sleep. If you fall into either of these categories, you are taking a positive step forward by reading this book and searching for some answers to help you get well. If, however, you don't exercise because you "cannot find the time" or you do not get enough sleep because of "too many obligations," then you may need to re-evaluate your priorities.

Some people try to rationalize reasons for not getting enough exercise and sleep. For example, they may claim to be "too busy trying to make that first million dollars" or the "second" or "third." If this is you, ask yourself how important money will be if you sacrifice your health in the process of making it. Or, you may tell yourself that exercise and sufficient sleep take too much time away from certain obligations that demonstrate your love for your family. Think of how much more love you will be able to give if your health is at its best. Is not the quality of love just as important as the quantity? How loving and how loveable can an unhealthy person be?

PROTEINS

Proteins are important for hormone and enzyme production, cell growth and repair, genetics and the manufacture of antibodies. In humans, proteins consist of 21 amino acids connected in chains. The sequence and length of the amino acid chains distinguish one type of protein from another. Protein can be attained by eating whole grains, beans, and eggs, as well as meats. Eggs are the most perfect proteins for humans.

Insufficient protein can affect the immune system as well as imbalance the body chemistry in general, due to its importance in the endocrine system, glands and hormones. Overcooking and microwaving of meats cause the protein in them to become inactive, while drinking liquids with meals disturbs protein digestion.

Mercury may cause an imbalance of protein in the body, inactivate sulfhydryl bonds found in some amino acids, and affect the protein of white blood cells that are responsible for immune responses. It is also capable of adversely affecting the protein components in cell replication by altering DNA molecules.

Mercury toxic patients may show improvement in protein metabolism with proper dietary protein intake, digestive enzymes, and free form amino acid supplements if properly combined with silver mercury filling removal.

CHOLESTEROL (FATS)

Contrary to popular belief, fats in the form of cholesterol are essential to life. The brain and nervous system is composed of a high percentage of cholesterol, and it is essential for hormone production. Indeed, you could not live without a certain amount of cholesterol.

Although a high level of cholesterol in the blood is believed to increase the chances of heart disease, this does not mean that the less cholesterol you have in your body the better off you are. There is an ideal level of cholesterol in the blood for optimum health, and levels below that ideal can put a strain on the endocrine glands and central nervous system. The ideal level of cholesterol may not be the same for everyone, however. For example, heavy exercise can significantly reduce the cholesterol levels. Smoking, along with the consumption of sugar, caffeine, and alcohol can increase cholesterol in the blood above the ideal level, while mercury toxicity is believed to decrease cholesterol below the ideal level.

In body chemistry balancing, the proper interpretation of the cholesterol level is essential. The level indicated from a blood test can be very deceptive in relation to the person's true condition of health. For instance, you may consume large amounts of caffeine which, by itself, would tend to increase the level of cholesterol. But if you also have a high exposure to mercury, there could be a corresponding decrease that may falsely bring your blood cholesterol level back to the ideal level. Two wrongs do not make a right, however, and in the above example, an unhealthy and potentially dangerous condition could remain undetected if all contributing factors are not considered.

Butter (not margarine) and eggs will help to optimize the blood cholesterol level once the above mentioned negative factors are eliminated. Although this viewpoint may, at first, seem contrary to the prevailing thoughts on blood cholesterol levels, butter and eggs are not necessarily a problem for persons with high cholesterol levels. The culprits are usually substances that interfere with metabolism, such as sugar, caffeine, alcohol, refined carbohydrates, and smoking, along with overweight and stress. The correction of high cholesterol is best achieved by the elimination of these negative factors rather than the elimination of all cholesterol foods.

SPECIFIC DIETARY RECOMMENDATIONS AND SUPPLEMENTATION SUGGESTED FOR MERCURY TOXICITY

The following recommendations are suggested for many, but not all, mercury toxic patients particularly during their initial stages of therapy. Depending on the results of follow-up laboratory tests and symptomology, these recommendations may be discontinued or modified after a few months, or they may be continued indefinitely. Since each person is different, a dietary and supplementation regime must be tailored to the individual. These suggestions are general guidelines and should be administered under the proper guidance of a qualified health care provider who is well informed on mercury toxicity and body chemistry therapy.

DIETARY

1. Vegetables fresh or steamed (1 to 2 cups per day)

2. Saturated fats (in the form of butter, 4 - 8 pats per day) and unsaturated fats (olive or safflower oil is suggested, 2 - 4 tsp. per day), consume these fats in the raw (unheated) state, if possible.

3. Fresh fruit (1 to 2 cups per day)

4. Eggs - boiled, poached, or scrambled (1 to 2 per day)

5. Whole grains - brown rice, oatmeal, millet, buckwheat, etc., but wheat is usually not recommended (1 to 2 cups per day steamed or cooked on low heat)

6. Meat - preferably turkey, lean beef, and game. Occasionally fresh (not frozen) organic chicken but pork is not recommended. It is suggested to obtain meats from animals that were

raised free of pesticides, antibiotics and hormones whenever possible. (Minimum of 4 ounces per day — game and beef should be cooked as rare as possible)

7. Beans, legumes, seeds, nuts (1 cup per day)

8. Eliminate all fish and seafood including seafood products and especially tuna. (One seafood meal per week can more than double the average mercury level in the blood.)

SUPPLEMENTS

Dosages vary with the individual and should be determined by a qualified health care provider based on laboratory tests and symptomology (Fig. 5-2). The following is a partial list of vitamins, mineral, herbs, and glandular supports that are often used in mercury toxicity.

1. Vitamin A

2. Vitamin C

3. Vitamin E

4. Amino Acids (particularly cysteine and methionine)

5. Folic Acid

6. Digestive Enzymes

7. Fatty Acids

8. Iodine

9. Zinc

10. Magnesium

11. Manganese

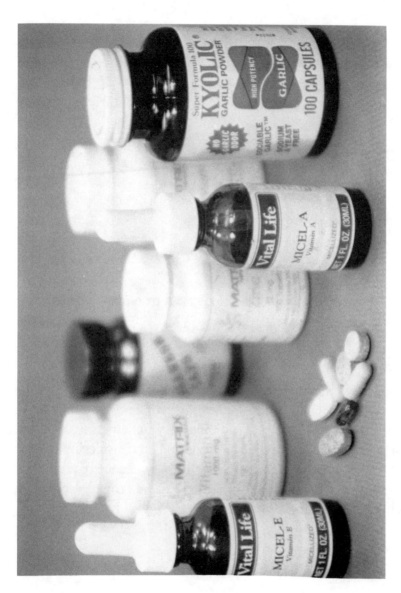

Fig. 5-2. Supplements: Various vitamins and minerals may help to mobilize and bind the stored mercury as well as balance the body's chemistry.

12. Potassium

13. Chromium

14. Sulfur

15. Selenium

16. Glutathione Peroxidase

17. B-Complex

18. Garlic or Kyolic

19. CoEnzyme Q_{10}

20. Germanium

21. Evening Primrose Oil

22. Homeopathy and Herbology can be of aid

23. Thyroid (low doses)

24. Pituitary (low doses)

In summary: Mercury from silver mercury fillings is only one possible factor that may contribute to unhealthy symptoms you may be having. Many symptoms associated with mercury toxicity are also associated with poor diet and unhealthy lifestyle. When considering mercury detoxification, you therefore, will have the best chance of recovery if you address your diet and lifestyle as well as the proper removal of your silver mercury fillings.

CHAPTER 6

Things To Know Before Searching Out a Dentist To Remove Your Silver Mercury Fillings

Before reading this chapter, be aware that these are difficult times for dentists to talk about mercury toxicity in relation to silver fillings. Mainstream dentistry has taken a firm stand opposing this concept and various organized dental groups have created some hardships for certain outspoken dentists.

Maria came for help after hearing me speak on a local talk show. Although she had many other health symptoms that were classic to mercury toxicity, her main complaint was general fatigue. She went through all the appropriate testing and evaluating, and progressed through the counseling sessions at a relatively fast rate of speed. Maria was very cooperative, enthusiastic, and she believed with all her heart and soul that this therapy could help her symptoms. I also believed that she could be helped and gave her the names of 3 dentists in her hometown that I personally knew and recommended. I suggested that Maria make consultation appointments with all three dentists and interview each one. Maria lived in a large metropolitan area, and I felt that she was very fortunate to be able to choose from several dentists. (Some patients must travel many miles to find a dentist that recognizes mercury as a potential problem.)

Maria interviewed those three dentists but she also interviewed Dr. M, her family dentist. Maria had been a patient of Dr. M's for 18 years and thought of him as "part of the family." She liked his personality and had trusted him with her dental care for most of her life. She did not want to change dentists, so she asked Dr. M to remove her silver mercury fillings.

Dr. M was probably a very nice man and a dedicated and sincere dentist; however, Dr. M was not well informed in the area of mercury toxicity. He tried to discourage Maria from going through with this therapy and told her that since he had originally placed all of her fillings, he could not bear the thought of her going to another dentist to have them all removed. So, he agreed to replace the silver mercury fillings if she would agree to use gold as the replacement filling material.

Maria agreed to allow Dr. M to do it his way.

There were 15 silver mercury fillings to be replaced, and on the first appointment three were removed. What do you think happened? Maria got sick! She felt nauseous while the fillings were being drilled out; and by the time she left Dr. M's office, she was physically and emotionally exhausted. She barely had the strength to drive home. At first she did not think too much about this because in the past she always felt exhausted after having silver mercury fillings replaced. She thought her symptoms were a result of the usual fear and anxiety that had always accompanied her visits to the dentist.Maria went home and "went to bed for 3 days." When she finally tried to get up again, she was sluggish and much weaker than she had remembered before the fillings were removed. Maria's second appointment was scheduled for three weeks later (it normally takes 2 or 3 weeks for a dental technician to fabricate gold fillings). On this appointment the three gold fillings were cemented into place and four more silver mercury fillings were removed. Guess what happened this time? Maria got sick again! She reported nausea and great fatigue while having the fillings removed. Yes, afterwards she spent more time in bed. She also "felt depressed for about 1 week."

On the third appointment, four more gold fillings were permanently placed and four more silver mercury fillings were removed. You, guessed it! Maria got sick again! More nausea, more fatigue, more time in bed, and more depression. Also, this time she developed a "headache that lasted 2 days." By now, Maria was desperate. She called me for a consultation and after telling her story, she asked me to call Dr. M and discuss her treatment with him.

Over the phone I could tell that Dr. M was skeptical and did not believe that the improper removal of silver mercury fillings could have anything to do with Maria's nausea, fatigue, depression, and headache. Dr. M resisted discussing the case with me.

I spoke to Dr. M on three occasions in the days that followed, each time meeting more and more resistance, until there was almost outright hostility coming from Dr. M. Finally I called Maria and suggested that she get a second opinion from one of the three dentists that were originally recommended to her.

Maria did finally see Dr. E, one of the recommended dentists. Dr. E was able to detect a very small amount of mercury filling material which had remained at the edge of one of Maria's gold fillings. He also discussed with her the possibility of exacerbation of symptoms when gold is placed in a mouth that has silver mercury fillings remaining. (See Chapter 3 under "Mercury Vapor Analyzer Screening" for a discussion of gold/mercury interaction.)

Maria continued her dental care with Dr. E, who eventually removed all of the seven gold fillings that Dr. M had placed. Of those seven gold fillings, four had residual mercury filling material remaining. Dr. M was acting in good faith. (Note that many dentists leave silver mercury filling material under gold fillings because that has been the "usual and customary" technique the dental schools have been teaching for years.)

Dr. E completed Maria's dentistry with relatively little aftershock. No more nausea or headache and only slight fatigue was reported. Maria is much better now and has become a highly vocal advocate for the proper treatment of mercury toxicity.

How can you prevent what happened to Maria from happening to you? And, what is involved with the proper removal of silver mercury fillings and treatment of mercury toxicity?

In the story above, Dr. E was successful in performing the dentistry for Maria without her becoming ill because he took several precautions. These precautions will be discussed in the remainder of this chapter along with other preventive measures related to mercury toxicity that even Dr. E did not know about at the time.

It is unlikely that you will find a dentist who incorporates all these items in his treatment of mercury toxicity; but here's what to look for:

HEALTH HISTORY QUESTIONNAIRE

An extensive health history and personal interview is the first and most important step in the treatment of mercury toxicity.

The health care provider should have you complete a rather lengthy questionnaire concerning your health symptoms. You will be asked many questions, sometimes more than once, to help jog your memory in relation to your symptoms. The questionnaire should include both general and specific physiological and dental related symptoms. "Have you ever felt ill immediately after having fillings placed?" might be one possible question. Some people will suddenly realize that a previous illness correlates with having dental treatment, but not until someone asks them to think about it.

TESTING FOR MERCURY VAPOR IN YOUR MOUTH

The Jerome Mercury Vapor Analyzer is thoroughly discussed in Chapter 3. It is an instrument which can tell you and your dentist how much mercury vapor is released by your fillings at any given moment.

At last count, there were only a handful of dentists in the world who had this machine. However, don't choose a dentist solely on the basis of whether or not he has one of these analyzers since there are many well-informed dentists who have not invested in this instrument.

TESTING THE ELECTRICAL POTENTIAL OF YOUR FILLINGS

Electrical measuring devices are also discussed in Chapter 3. These instruments measure the amount of electricity being generated by your fillings. Some fillings are said to generate negative current, while others have positive current. Negative electrical current may have the potential of producing the highest levels of mercury vapor from your fillings. Also, the higher the current, the greater the chance of high levels of mercury vapor. Some clinical researchers suggest that nerve related symptoms (multiple sclerosis, headache, depression, etc.) are most likely to show *high negative* readings from their silver mercury fillings. Symptoms related to the immune system tend to show a predominance of *positive* electrical currents. Since many dentists believe that the best results are obtained when fillings are removed in a particular sequence according to their positive or negative charges, it may be wise to choose a dentist who has one of the electrical measuring devices to follow through with this aspect of therapy.

BODY CHEMISTRY BALANCING

Since mercury is a poison that has a detrimental effect on the protoplasm of all living cells, it is easy to suspect mercury poisoning whenever facing undiagnosable, untreatable health symptoms related to any of the enzyme or endocrine systems in the body. Mercury's negative effect on the endocrine glands (thyroid, pituitary, pancreas, adrenals, etc.) may contribute to poor digestion, absorption, and assimilation of foods. In treating for mercury toxicity, then, an evaluation of the body chemistry is recommended. Balancing the body chemistry involves evaluating and monitoring blood, hair, and urine tests. It is based on nutrition therapies, including vitamin and mineral supplements, endocrine supplementation and lifestyle modifications. Some dentists involved with mercury toxicity have become very well versed in these areas. Other dentists who do not perform testing and counseling themselves, can refer mercury toxic patients to nutritionists, physicians, chiropractors, and naturopaths for evaluation and therapy. It is important that the practitioner be aware that there are specific areas to be addressed in balancing the chemistry of the mercury toxic patient.

When considering therapy for mercury toxicity, balancing body chemistry may be just as important as the actual clinical dentistry of silver mercury filling removal.

The mercury released from your fillings may be the biggest source of mercury contamination to your body, but removing your silver mercury fillings may only prevent further chronic exposure to mercury. Removing the mercury from the tissues of the rest of your body and aiding in the correction of its side effects are the critical factors in the success of your therapy. Body chemistry balancing through nutrition, supplementation, and lifestyle modification are of particular importance for most individuals. More detail about body chemistry balancing is found in Chapters 3 and 5.

Find a dentist who is concerned about your body as a whole, as well as your teeth to achieve the full potential of improvement in your health.

MAGNIFYING GLASSES

When it comes to detoxifying mercury from your body, it is absolutely imperative that every last minute particle be removed from your teeth. Magnifying glasses greatly aid in the task (Fig. 6-1).

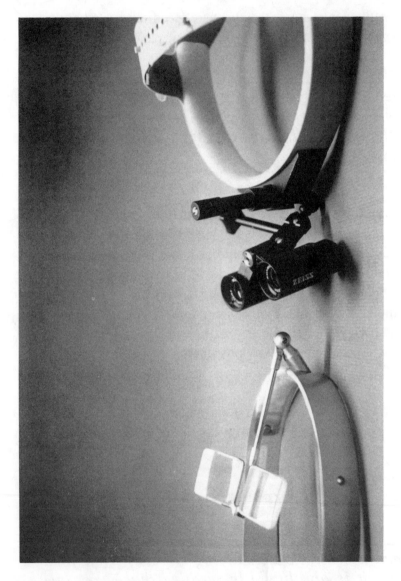

Fig. 6-1. Magnifying Glasses (Loupes): Used by dentists who understand the importance of removing every particle of silver mercury filling material and decayed tooth structure. (Photograph courtesy of Carl Zeiss, Inc.)

It is difficult to remove all the mercury filling from a tooth in a conservative manner without magnification.

The special glasses allow less tooth structure to be disturbed because they make it easier for the dentist to determine precisely where the filling stops and the tooth begins.

FIBEROPTIC LIGHT FROM THE HANDPIECE (DRILL)

Everything that I have indicated about magnifying glasses also applies to *fiberoptic lights* on the dentist's drill (Fig. 6-2).

Practically every dentist has an overhead light that is usually attached to the ceiling or a pole beside the chair, but not every dentist has lighting coming out of the drill. If the dentist doesn't have additional light, he may not be able to adequately see inside the deepest part of the tooth. This is where the lighting is particularly important because small particles of mercury filling can be left behind.

Some dentists wear a special light attached to a headband which gives additional lighting but is not as good as fiberoptic light coming out of the drill and shining directly into the tooth cavity.

Ask your dentist to show you the type of lighting he uses.

PATIENT PROTECTION WHILE REMOVING SILVER MERCURY FILLINGS

When the dentist is drilling out the silver mercury fillings, there will be very high levels of mercury vapor being released. No one has yet dis-

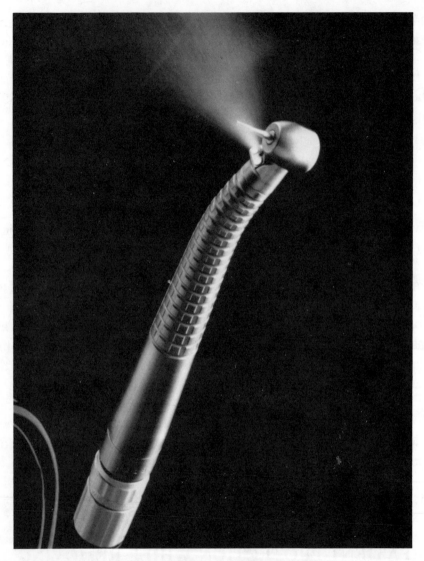

Fig. 6-2. Dentist's Handpiece (Drill): Notice the fiberoptic light that is being emitted from the head of this handpiece. The special lighting allows the dentist better visibility into the dark cavity of a tooth. This reduces the possibility that small particles of silver mercury filling material will be undetected and left behind. (Photograph courtesy of Midwest Dental Division of Sybron Corporation).

covered a way to completely protect the patient from absorbing some of this newly formed mercury vapor. However, steps can be taken to minimize exposure.

Rubber Dam

A rubber dam is a thin sheet of rubber that covers the mouth but allows only the tooth or teeth to be uncovered (Fig. 6-3). When the silver mercury fillings are being drilled out, the dental assistant uses a high volume suction to quickly remove the particles of silver mercury fillings and much of the mercury vapor. The rubber dam also helps to prevent silver mercury filling particles from falling back into your mouth and being swallowed.

Oxygen or Compressed Air

Some of the mercury vapor generated from the drill may not be stopped by the rubber dam and may not go into the high volume suction. Breathing oxygen or compressed air through a nose piece will further help diminish your exposure (Fig. 6-4). Most dentists have oxygen available in their offices for use with laughing gas or for an emergency situation. One word of caution, however: some people should not breathe pure oxygen for extended periods of time. For these people it is usually best to breathe compressed air, or as an alternative, they may breath oxygen only during the times when the silver mercury fillings are actually being drilled out and for a few minutes afterward until the high levels of mercury vapor have been diluted by the air in the general vicinity.

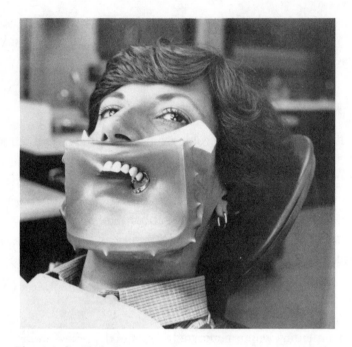

Fig. 6-3. Rubber Dam: Used during removal procedures to prevent accidental swallowing of small particles of silver mercury filling material as it is removed. (Photograph courtesy of The Hygienic Corporation).

Fans and Ventilation

I recommend opening windows in the room where mercury fillings are being removed to diffuse the concentrated mercury vapor in the immediate area. A small portable fan that creates a draft close to the patient's mouth and nose is also a good idea (Fig. 6-5).

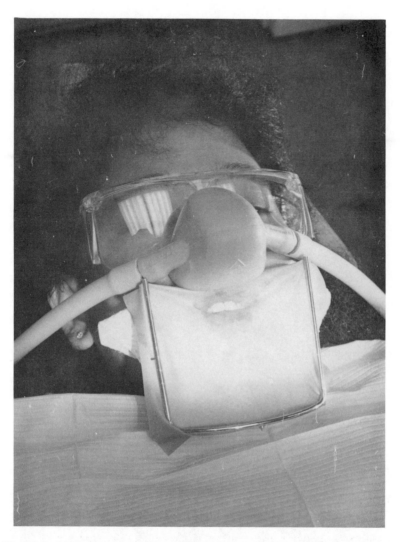

Fig. 6-4. More Patient Protection: The administration of oxygen or compressed air through a nosepiece helps prevent the inhalation of high mercury vapor levels that are released when the silver fillings are removed. Protective glasses reduce the chances of small particles getting into the patient's eyes during the drilling procedures.

Fig. 6-5. Ventilation: Having a nearby fan blowing fresh air (from an open window) over the patient's mouth and face helps to dilute the very high levels of mercury vapor that is release by the dentist's drill.

Protective Coverings

In spite of good intentions, mercury filling particles can find their way onto a patient's body and clothes and even into their eyes. Draping the patient's entire body except, of course, the face, with a large plastic apron designed for this purpose is helpful (Fig. 6-6). Also, protective glasses are recommended, and patients are encouraged to keep their eyes closed whenever the dentist or assistant is working in their mouth. Draping the patient's hair or covering the head with a plastic shower cap is also a good idea.

INTRAVENOUS VITAMIN C

Mercury vapor is almost impossible to stop. It is said to "walk through walls." No matter what you and the dentist do, some mercury will probably penetrate your body. Vitamin C helps chelate (bind) mercury, and I suggest that a slow intravenous drip of diluted ascorbic acid (Vitamin C) be given while silver mercury fillings are being removed (Fig. 6-7). The Vitamin C will bind the mercury that finds its way into your blood stream and force it to be excreted through the urine. Without intravenous Vitamin C during mercury filling removal, some patients are very likely to get ill for several days, or even weeks, afterwards. Ask that your dentist provide you with intravenous Vitamin C, either by himself, or by another licensed or qualified health care provider.

The suggested dose for adults is 50 to 60 grams with 400 cc of saline. Some patients, however, cannot tolerate this much in the beginning, so it is sometimes necessary to begin with a lesser dose.

Fig. 6-6. Patient Drape vs. Patient Napkin: During dental procedures a large drape (left), that completely covers the patient's body, affords considerably more protection than the relatively small standard napkin (right).

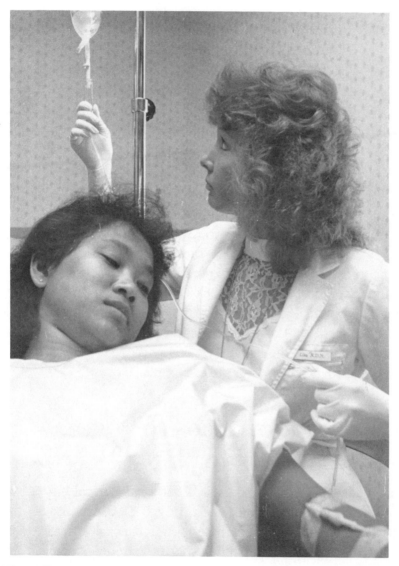

Fig. 6-7. Intravenous Vitamin C: Some mercury may be absorbed by the patient during the removal process in spite of adequate patient protection. A slow continuous I.V. drip of Vitamin C will help reduce post operative symptomology due to the chelating ability of Vitamin C on mercury.

X-RAYS

I discourage the indiscriminate use of X-rays but feel that sometimes they are a necessity. If you have only a few very small silver mercury fillings, the dentist may not require x-rays. However, most people's fillings do not fall within this category. Some teeth with root canal treatment can only be properly assessed using an x-ray. There are also large, deep fillings near nerves, fillings under the gums, and abscesses at the roots of teeth which cannot be easily detected. Requirements for medical, legal and dental insurance must also be considered. There are many justifications for using x-rays, and in the situation of silver mercury filling removal, they may be necessary. A full mouth survey of 15 or 20 x-rays is not always required. I recommend taking at least one panoramic x-ray showing the entire mouth before filling removal. This is a minimum for most cases; but, occasionally, other small x-rays may be necessary to get a clear picture of an unusual situation that cannot be determined otherwise. Also be sure that the dentist or dental assistant drapes your body with the lead apron before they take any x-rays. A lead collar that covers the neck is also advisable.

SELF PROTECTION FOR THE DENTIST AND ASSISTANT

I recommend that the dentist and the assistant wear surgical gloves, protective glasses, and mercury filter masks (Fig. 6-8). Gloves are of obvious importance with the high incidence of communicable diseases today such as AIDS, herpes, and hepatitis. Consult your dentist concerning this area of protection for themselves.

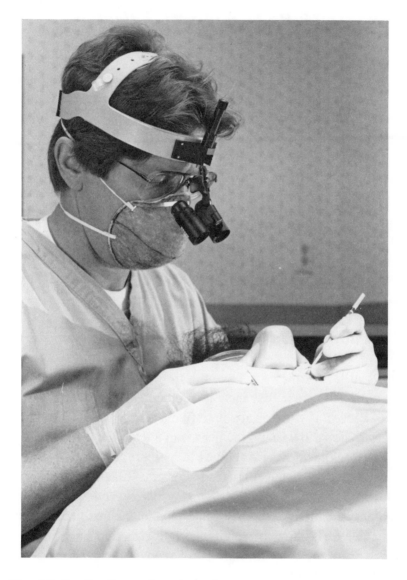

Fig. 6-8. Dentist Protection: A dentist may wear a special mercury-vapor-filtering-MASK, scrub clothing and protective glasses. Gloves are worn for both doctor and patient protection.

OTHER HEALTH CARE PROVIDERS

It is essential that any health care provider understand the benefits and limitations of his own ability and realize the absolute necessity of using experts in other fields. Following is a partial list of types of health care providers that may be utilized in helping the patient with mercury toxicity.

Dental Specialists

Some dentists are particularly well-trained in full mouth reconstruction and jaw joint problems (TMJ). They are not necessarily the same dentists who test and counsel you or even remove your fillings. You may happen to find a "TMJ" dentist who is also well versed in mercury toxicity.

M.D.'s and D.O.'s

M.D.'s and D.O.'s may not have the time to become well versed in mercury toxicity since they have to keep abreast of so many other aspects of illness. If you do happen to find a doctor who is familiar with mercury toxicity, it will aid the dentist in a team approach concerning your health. Some dentists involved in mercury toxicity will refer you to one of these physicians to make the original diagnosis. Or they may be called upon to help with EDTA and Vitamin C chelation as well as other detoxification therapies. Some physicians can provide "sweat therapy" which is a logical step in the right direction since poisons are removed from the body through all of the excretory organs (kidney, liver, lungs, skin, etc.).

Nutritionists

Too much emphasis cannot be placed on the importance of nutrition, supplementation and lifestyle modifications regarding the treatment of mercury toxicity. A qualified nutritionist is vital to the success of therapy. By "qualified," I mean someone who is familiar with the specialized needs of patients suffering from mercury toxicity or willing to work with the dentist in this area. I believe that each person's diet should be tailored to his specific nutritional needs which have been determined by evaluation and monitoring of various blood, hair, and urine tests along with health histories and symptomology. Mercury's effects may be seen in these tests, but only by someone specially trained in mercury toxicity therapy.

There are nutritionists in private practice as well as those working out of dentists', physicians', and chiropractors' offices. Chapter 5 goes into greater detail concerning the role of nutrition in the treatment of mercury toxicity.

Chiropractors

In recent years chiropractors and dentists have started to utilize each other's services. There are areas of overlap between the two, particularly in nutrition and supplementation, as well as jaw joint problems (TMJ syndrome). Since most dentists are not set-up for testing and evaluating nutritional needs, referring patients to a qualified chiropractor is sometimes very convenient.

Of particular interest with regard to chiropractic is that some doctors can perform adjustments that are said to "open the lymphatics for increased drainage and detoxification."

Also, some chiropractors have told me that their adjustments will last longer, in many cases, after a patient has had their silver mercury fillings removed. There are several possible explanations for this phenomenon. First, removing a patient's fillings changes the bite and plays a role in the function of the TMJ. Secondly, this joint is intimately involved with the vertebral column and muscles of the neck which may relax as the TMJ and teeth become aligned. Also, removing the fillings removes the electrical currents that are being generated and which may have been exerting an effect on the nervous system and spinal cord, particularly in the neck region.

It will be interesting to see how chiropractors and dentists work together in the years to come as the treatment for mercury toxicity unfolds.

Psychotherapists

Mercury has a tremendous toxic effect on the nervous system, which can create or exacerbate emotional symptoms. Since dentists are not particularly trained to handle such areas as depression, anxiety and suicidal thoughts, orthomolecular psychotherapists are sometimes called upon.

In regard to mercury toxicity, one unusual emotional problem must occasionally be addressed. That is the problem that develops when the patient gets considerably better and various family members do not know how to adjust to a sometimes drastic change in circumstances. The best results follow when this issue is addressed before the patient's mercury toxicity therapy begins. Family therapy may sometimes be necessary to resolve some of the patterns that the family may have developed around the illness and how to cope with changes and recovery.

Accupressurists

There has been a growing interest in accupressure among some dentists in the United States, as well as some European countries, partially in relation to the mercury toxicity problem. Accupressure is a method of evaluating the status of various organ systems situated along the body's meridians. (Meridians are invisible energy pathways that follow various lines through the body.) Since most of our meridians pass through or near our teeth, a poisonous silver mercury filling in a tooth that lies on a certain meridian can cause an energy disturbance within an organ system that is associated with that same meridian. Consequently, the function of the organ may be impaired. This may seem unusual to those who have not studied accupressure, but it is the only explanation that I have for the following patient experience.

Bob came to see me for pain in one of his teeth. The tooth had a crown covering a silver mercury filling which, in turn, covered a metal post that was part of a root canal treatment. This tooth had been through practically every treatment imaginable, but it still bothered him. He had seen several dentists concerning this tooth over a two-year period. Each dentist tried something different, but the problem continued. I decided that mother nature was trying to tell Bob something about that tooth, and suggested that the tooth be removed. He agreed. While waiting for the tooth to completely anesthetize, he happened to mention that the only time a chronic knotting in him stomach seemed to ease up was when the tooth was numbed. I asked him how long he had had the knotting stomach problem and his answer was, "about two years." The tooth problem seemed to start at about the same time as the stomach knot, and vice versa. He also said that he had been to several physicians about the stomach knot, but none of the drugs or therapies prescribed ever helped him for very long. Immediately upon removing the tooth, Bob blurted out, "The stomach knot suddenly disappeared completely." One year later, the problem had not returned.

Experiences similar to Bob's may happen more often than we realize in dentistry. Many times patients do not report these experiences to their dentists because they fear their dentists will think they have lost their minds. However, dentists who know about accupressure and its effects are usually informed about mercury toxicity as well. They have heard many similar experiences from their patients concerning the invisible connection between teeth and organs of the body.

Don't forget about accupressure as an adjunct to mercury toxicity therapy.

Homeopathic Health Care Providers

Homeopathic treatment uses non-drug, non-toxic, highly diluted remedies that are designed to stimulate your body to heal itself. This form of treatment performs much differently than medicines and drugs. All drugs, even aspirin, have side effects. So, at best, drugs give only temporary relief by covering up symptoms and masking the real causes of the ailment.

Most homeopathic dentists are opposed to the use of mercury; however, not all dentists who are well qualified in treating mercury toxicity are necessarily well versed in the use of homeopathic remedies, and vice versa.

Although your treatment may not require homeopathics, it would be in your best interest to find a dentist who discourages the use of drugs and, if necessary, can refer you to a homeopathic health care provider who is knowledgeable about treating mercury toxicity.

Some homeopathic health care providers believe that homeopathic remedies for mercury toxicity work best after the silver mercury fillings have been removed; but others feel that remedies enhance the detoxification of the body's burden of mercury if used during the dental procedures.

INTESTINAL FLUSHES

While the dentist removes a patient's silver mercury fillings, small particles of filling material may be inadvertently swallowed by the patient. These particles usually pass through the body within one to three days or longer if the person has digestive problems and bowel constrictions. Keep in mind that the longer it takes the particles to pass, the greater the chance of mercury being absorbed through the intestines. Also, there are bacteria in the intestines that may change elemental mercury into organic mercury (methylmercury) which is much more toxic. Some research has shown that more mercury is excreted through the intestines by way of the liver and bile than by any other excretory organ. It is important to help your intestines cleanse the poisons as quickly as possible and to keep the bowels moving, especially during mercury toxicity therapy.

There are several things you can do to assist your intestines as mercury passes through them.

First, be sure that your bowels are functioning properly. If you are not having at least one easy bowel movement per day, you may have a problem. A diet that contains whole grains (brown rice, oatmeal, millet, rye, barley, etc.) two to three times per day and plenty of raw vegetables and fruits, should get your bowels moving. A nutritionist and/or a colon doctor should be able to help you if you are constipated.

Secondly, you may want to "flush" your intestines after each dental appointment involving work on your silver mercury fillings. If possible, this should be done immediately after leaving the dental office. There are many ways to "flush" the intestines, but I suggest using Vitamin C when removing silver mercury fillings because of the added benefit of Vitamin C's ability to bind mercury. Here is how the flush is performed:

Vitamin C Intestinal Flush

Swallow 1 tsp. of sodium ascorbate powder (Vitamin C) dissolved in 4 ozs. of water or diluted fruit juice. Repeat every 15 minutes until you experience diarrhea. For most people it takes 6-12 tsp. or 1-1/2 to 3 hours before they get results. If you have no diarrhea after 16 tsp., discontinue and start over the following day. If you get stomach cramps soon after taking a dose of Vitamin C, cut the amount in half (1/2 tsp. instead of 1 tsp.) for subsequent doses. Once diarrhea has occurred, discontinue taking the Vitamin C for this particular intestinal flush.

Note: There is more than one kind of Vitamin C. Be sure to use sodium ascorbate powder only. The other Vitamin C supplements will not produce the desired results for an intestinal flush and calcium ascorbate will contribute to an imbalance in calcium and magnesium for some mercury toxic patients.

Sodium ascorbate powder is difficult to find but can be special ordered through your nutritionist or vitamin/mineral specialty store.

Within the intestines are many beneficial bacteria. During digestion they help us assimilate nutrients. An intestinal flush may partially eliminate those bacteria, so it is wise to replace them. Here is one way to do it:

Repopulating the Gut With Beneficial Bacteria

Three times daily for one month, take 1/2 tsp. of lactobacillus acidophilus culture dissolved in 2 ozs. of water or fruit juice. Lactobacillus acidophilus can be obtained from your nutritionist or vitamin/mineral specialty stores.

Liver and Gall Bladder Flush

A liver and gall bladder flush is sometimes suggested one week after the removal of all silver mercury fillings. Your nutritionist can help you with this procedure. Anyone with liver problems or difficulty in assimilating fats, should be cautioned against liver flushes. The liver removes poisons from the blood and dumps them into the gall bladder. The gall bladder, in turn, dumps these toxic substances into the small intestines along with the bile. The poisons can then be reabsorbed along with the digestion and absorption of food thus "recycling" through the body. The liver and gall bladder flush allows you to excrete mercury that is being recycled (not to mention ridding your body of stored gall stones and other poisons in the gall bladder).

Caution — Liver/gall bladder and intestinal flushes should be supervised by a qualified professional.

* * * * *

One final note before concluding this chapter: As with any dental procedure, there is always a possibility of complications if you should decide to have your silver mercury fillings removed. For example, tooth sensitivities have been reported after the replacement of some fillings. The sensitivity usually diminishes after a short while; but if it continues, further treatment may be necessary. In some cases, tooth sensitivity can become more severe, requiring temporary sedation, endodontic therapy (root canal treatment), and even tooth extraction. The chance of filling and/or tooth fractures sometimes increases as the restorations become larger. Also, TMJ (jaw joint) symptoms can occur. Be certain to discuss with your dentist all of the possible complications that could arise before deciding to have your silver mercury fillings removed.

CHAPTER 7

Where To Start Looking
for the Right Dentist

I t is estimated that several thousand dentists throughout the United States are opposed to the use of mercury in dental fillings. Of those, only a few hundred have an adequate amount of special training to care for patients seeking help for mercury toxicity. That brings us to an average of, perhaps, four dentists per state, so you may have a difficult search ahead in finding a dentist who can give you proper treatment for mercury toxicity. One reason for my having written this book is out of concern for the people in need who do not know where to turn. If you are willing to devote your time and effort to doing some homework, you should be able to find the dentists nearest you who are familiar enough with mercury toxicity to help you. I recommend that you continue your search until you find several dentists from whom to choose. Understand that even if you find ten dentists who are helping mercury toxic people, each of those ten dentists is at a different level in his knowledge and experience.

First, ask a physician who specializes in *environmental medicine* for a referral to a "mercury-free" dentist in your area. Or, you may want to ask for the name of the dentist of your friends and associates if any of them have already had their silver mercury fillings removed. You might also check with the manager of your local health food store. If you know an open-minded M.D., D.O., nutritionist, chiropractor, or accupressurist, ask them for a referral. Health minded professionals usually run in the same circles and know each other or have, at least, heard of one another.

Also, many dentists opposing mercury belong to one or more of the following organizations. It is with these organizations, that dentists get most of their training on mercury toxicity. I suggest that you write these organizations and ask for names and addresses of their members who practice near you. These organizations may also provide you with additional information on mercury toxicity.

The International Academy of Oral Medicine and Toxicology (IAOMT)
550 - 6th Ave, S.W. Suite 950
Calgary, Alberta
Canada T2P0S2

The *IAOMT* sponsors conventions and provides constantly updated scientific data to dentists concerned with the mercury issue.

The Toxic Element Research Foundation (TERF)
P.O. Box 80
Colorado Springs, CO 80901

TERF periodically offers three to six day seminars to dentists and other health care providers on mercury toxicity. It conducts research in the area of heavy metal toxicity and offers a computerized nutrition/ mercury toxicity evaluation service for persons who do not have a qualified dentist in their immediate area.

Health Consciousness
P.O. Box 550
Oviedo, Florida 32765

Health Consciousness is a trade magazine edited and published by a physician, Roy Kupsinel, M.D., who himself was treated for mercury toxicity. This magazine helps to keep dentists and other health care providers informed on the issues of mercury toxicity as well as holistic health.

The Mittelman Newsletter
263 West End Ave., #2A
New York, New York 10023

Many dentists subscribe to this dental newsletter. The editor/publisher, Jerry Mittelman, D.D.S., will send you the names and addresses of the mercury-free dentists that subscribe to this publication in your area.

Bio-Probe Newsletter
P.O. Box 580160
Orlando, Florida 32858-0160

Probably more mercury-free dentists subscribe to the Bio-Probe Newsletter than any other publication that addresses mercury toxicity.

The International Association of Electro-Accupressure
According to Voll-Dental Chapter
1441 Kapiolana Blvd. Suite 721
Honolulu, Hawaii 96814

This is an organization of dentists who study accupressure in dentistry. Any dentist who incorporates dental accupressure in his practice will most likely be opposed to silver mercury fillings.

National Center for Homeopathy
1500 Massachusetts Ave. N.W., Suite 41
Washington, D.C. 20005

Any dentist who uses homeopathy in his practice is probably not placing silver mercury fillings.

Holistic Dental Association
974 N. 21st Street
Newark, Ohio   43055-2922

Dentists who think of themselves as being holistic are generally well informed about the toxic effects of mercury.

The Society of Ultramolecular Medicine
6125 West Tropicana
Las Vegas, Nevada   89103

This organization is comprised of physicians and dentists interested in homeopathy, acupuncture, and electrodiagnosis. It provides a means for exchanging ideas and discovering new techniques involving the bioenergetic circulatory system of the body.

*   *   *   *   *

If you discover a dentist who belongs to several of the above organizations, especially if he is active with the *IAOMT* and *TERF*, you can be sure that he is worthy of your consideration. He is a well-informed health care provider who is concerned about the effects of mercury as well as other poisons.

CHAPTER 8

Interviewing the Dentist

Once you find the dentists in your area who are knowledgeable about mercury toxicity, you will still need to determine which one can help you most. Call the office of each dentist on your list and ask for a 30-minute to 1 hour consultation appointment. You may encounter some resistance from the receptionist. She doesn't generally have patients call and ask for a consultation without x-rays and an examination. If she is uncooperative, explain that you are considering having all your silver mercury fillings removed and you are talking with more than one dentist before deciding which dentist to choose. Even though you do not want an examination or x-rays at this appointment, be prepared to pay the dentist for his time.

Don't be surprised if the receptionist puts you on hold. She is probably checking with the dentist to see if she can reserve 30 minutes on his schedule for a consultation and if so, what the fee would be. If possible, don't schedule an examination at this point. It will take up too much of the dentist's time, and usually you will not be able to talk with him and ask the questions necessary to help you make an informed decision.

Sometimes you may have to agree to an examination; but you don't have to have x-rays taken more than once, especially if you provide your own set. It's a good idea to arrange for obtaining a duplicate set of x-rays from the first office you visit. Most offices can accommodate this request; and you can then take your set from office to office. Just be sure to make your request before any x-rays are taken. The type of x-rays necessary are discussed in Chapter 6.

Some offices will have a consultation appointment after an initial examination appointment. In this case, they may call it a "case presentation" or "a treatment plan consultation." You will need this later, of course, but for now all you want is time to talk to the dentist and interview him.

At the consultation, begin by explaining that you are looking for a dentist to replace the silver mercury fillings in your mouth. Don't go into the details of your health symptoms at this time.

Remember, you want to interview the dentist, not he interview you.

Your purpose is to gather as much information as possible about the dentist being interviewed. Most dentists will cooperate once they understand your purpose. Ask the dentists if they have read this book. If they have not, show them this chapter on "interviewing." If they still won't answer your questions, strike them off your list and interview the other dentists.

Realize that you do have the right to know about a dentist's skills, experience, and knowledge before entrusting him with your health.

Following is a set of questions that will give you a good basis for evaluating a dentist's ability to handle the mercury toxic patient. I have included a paragraph on each question, but to keep from repeating myself, I may refer you to other sections of the book for more information. You are encouraged to ask other questions as you become more knowledgeable in the area of mercury toxicity. All the questions have been listed together at the end of this chapter.

1. *How long have you been opposed to the use of mercury in dentistry?*

Although this first question isn't really necessary, it will "break the ice." One dentist may confess that he only recently began to take this issue seriously; that he has just returned from a 3-day seminar on mercury toxicity. This dentist may be worth considering, because he is excited and enthusiastic and he may go out of his way to do everything he can to help you. Another dentist may say that he has been opposed to mercury for over ten years; however, he may still be using it. His reason may be that most of his patients cannot afford gold or porcelain and he doesn't think that plastic (composite) is strong enough to last when used on back teeth.

Whatever his reason, reject any dentist who is still using mercury.

The majority of dentists treating mercury toxicity have been heavily involved with it for only a few years. I know of only three dentists who have studied mercury toxicity for more than ten years and only one who has never placed a silver mercury filling in his entire career of over 20 years. These dentists are the exception; so it is highly unlikely that the average person will run into one of them.

If you interview a dentist who has been actively studying and practicing mercury-free dentistry for three or four years, you have found a veteran in this area.

2. *How did you get interested in mercury?*

Many of the dentists opposed to mercury in dentistry have themselves been treated for toxic symptoms. Sometimes a dentist will become interested because of a dramatic experience of a friend or loved one who had mercury toxic symptoms. These dentists are usually more than happy to relay these experiences to you.

3. *About how many mercury toxic patients have you helped by removing their silver mercury fillings?*

If a dentist is involved with only the actual removal and placement of silver mercury fillings, and refers his patients to other health care providers for detoxification procedures, he is probably spending an average of 8 hours per patient. If the dentist spent all his time treating only mercury toxic patients (which is almost unheard of) only about five new patients per week could be seen. This adds up to approximately 250 patients per year. If the dentist has done this type of work for four years (rare, indeed), he has seen approximately 1,000 patients. However, this is probably not the case at all. I have found that I may spend an average of 20 hours per patient, with testing, consultations, evaluations, detoxification, follow-up testing, and actual dentistry. Add to that the fact that most general dentists are seeing patients other than those with mercury toxic symptoms. These factors alone would suggest that on the average a dentist could not see more than one new mercury toxic patient per week. Simple mathematics will give you 50 patients per year, and after 4 years the dentist will have seen about 200 mercury toxic patients. Even 200 may be an exaggeration. If you find a dentist who has been involved with mercury toxicity for 4 years and has seen over 50 mercury toxic patients, you have found one of the most experienced dentists in mercury toxicity.

Beware of any dentist who claims to have treated thousands or even several hundred patients for mercury toxicity.

4. *How did you get your knowledge concerning mercury toxicity?*

Until the early 1980's there were no journals, seminars, or conventions available where a dentist could learn about mercury toxicity. There were a very small number of dentists throughout the world who were researching and practicing on an individual basis without very much organization between themselves. In 1983 the *Toxic Element Research Foundation (TERF)* was formed by Dr. Hal Huggins of Colorado Springs, Colorado. *TERF* began to offer one- and two-day seminars on mercury toxicity. At about the same time, Victor Penzer, D.M.D., M.D., along with the late Theodore Ingalls, M.D., began to offer two-day seminars on mercury intoxication out of Boston University. These have, so far, been the only postgraduate courses available on a regular basis for interested health care providers.

Another organization that attracts some of the most conscientious dentists involved with mercury toxicity is called *The International Academy of Oral Medicine and Toxicology (IAOMT)*. It was formed in 1984, and presently has about 120 members throughout the world. The main purpose of this organization is to keep dentists informed of new scientific information concerning mercury toxicity. The *IAOMT* has a monthly journal as well as periodic conventions dealing with mercury and other dental poisons. The conventions feature speakers from all over the world who are considered to be the most brilliant scientists involved in heavy metal toxicity.

Other organizations that may have introduced a dentist to mercury toxicity are the *Holistic Dental Association* and various accupressure and acupuncture groups.

5. *How do you keep your knowledge current concerning mercury toxicity?*

Your purpose for asking this question is, of course, to determine if the dentist understands that this issue of mercury is in its infant stages and that new knowledge is discovered daily.

The answers to this question may overlap with the preceding question #4. The dentist can keep current on new developments by remaining active with the *Toxic Element Research Foundation* and the *International Academy of Oral Medicine and Toxicology*. Both *TERF* and

IAOMT have periodic newsletters called *"Momentum"* and *"Bioprobe"* respectively. There are also publications dealing with mercury toxicity such as *Health Consciousness* and the *Mittelman Letter*. The dentist may also get information from the *Holistic Dental Association* and organizations involved with accupressure/acupuncture. (See Chapter 7 for more information on the above organizations and newsletters.) I know of one "support group" of dentists and other health care providers in Denver, Colorado, which at one time, was meeting on a monthly basis to discuss personal experience with treating mercury toxic patients.

Look for a dentist who demonstrates willingness to keep abreast of new developments in the area of mercury toxicity.

6. *What things do you do to test and monitor for mercury intoxication?*

Some dentists, but not all, refer their patients to a physician for the initial testing and evaluating for mercury intoxification. As a minimum, blood, hair, and urine mercury testing may be performed and evaluated well before the silver mercury fillings are removed. These same tests may be performed a few weeks after all the fillings are removed, and periodically as months go by. Sometimes the dentists will order these tests, especially if they have had the training. Occasionally, the testing and monitoring will be turned over to a nutritionist, chiropractor or some other health care provider. However, understand that these other health care providers are rarely trained in the area of mercury toxicity. Proper interpretation of body chemistry through blood, hair, and urine testing is complicated and extremely specific. Just because a health care provider is familiar with blood, hair, and urine tests, doesn't mean that he is familiar with the area of mercury toxicity. More information concerning blood, hair, and urine tests can be found in Chapter 3.

Accupressure, kinesiology and homeopathy can be invaluable in the hands of properly skilled health care providers. But, these are disciplines usually recommended as an adjunct to blood, hair, and urine tests when it comes to evaluating the mercury toxic patient.

7. *In what ways do you help a mercury toxic patient to detoxify the mercury that has been absorbed into the tissues of the body?*

How a dentist answers this question will give you insight into whether or not he understands that mercury toxicity is much more complicated than just removing silver mercury fillings. It is true that the fillings are usually the biggest source of exposure to a patient, but they are not the only source. Also, just removing fillings with no other form of therapy is probably not going to help very many people with mercury toxicity.

The real help comes with the removal of the mercury that has contaminated tissues of various parts of the body. This is accomplished mainly with nutrition, supplementation, and life style modifications which are topics covered in depth in Chapter 5. In general, the dentist or other health care provider should counsel you on these additional aspects to help mobilize and excrete mercury that has accumulated in and on the cells of your body tissues. Other sources of mercury must also be identified and eliminated. For example, you may be working around certain pesticides or using certain cosmetics, either of which may contain mercury. Just removing your fillings isn't likely to help you very much in these cases. Also unhealthy aspects of your lifestyle may be hampering your body's ability to excrete mercury. For example, smoking, alcohol, caffeine, sugar, and refined carbohydrates all take their toll on the optimum functioning of your body. Accupressure, homeopathy, chiropractic, chelation, and colonic therapy may also provide additional benefits in mercury detoxification.

Look for a dentist who can augment his dentistry with nutrition, supplementation, and lifestyle modifications as well as other alternative therapies to enhance your body's ability to excrete mercury from its cells.

8. *What other health care providers do you work with in relation to helping people with mercury toxicity?*

A conscientious dentist will have several other health care providers with whom he consults and refers when the need arises. Physicians may make the initial diagnosis and will sometimes participate in the detoxification process. Nutritionists may be called upon if the dentist does not, himself, have time to test, evaluate, and counsel. Chiropractors may take care of nutritional and supplemental aspects as well as provide some chiropractic adjustments to enhance detoxification. Nutritionists and accupressurists as well as colonic and massage therapy are sometimes used. Other dentists who are specialists in particular areas of dentistry may also be necessary. More information concerning other health care providers is covered in Chapter 6.

Look for a dentist who accepts the limitations of his own expertise and who is willing to refer to other health care providers.

9. *Do you remove fillings in a particular order?*

Many clinical researchers have noted the importance of sequential removal. It only takes a dentist a few minutes to take electrical readings of silver fillings. So you may want to have it done as an added precaution. More information on electrical readings and sequential removal of fillings can be found in Chapters 3 and 6 of this book.

10. *What precautions would you take to protect me from further mercury exposure from vapor and particles released while removing my fillings?*

Whenever a dentist removes a silver mercury filling, extremely high levels of mercury vapor are released. I have recorded as much as 600 mcg/m^3 of air in a patient's mouth immediately after a silver mercury filling was removed. As mentioned previously, (see Chapter 6 under *Mercury Vapor Analyzer Screening*), this is many times higher than the *OSHA*'s maximum safe limit of fifty (50) mcg/m^3 (for the work area) and considerably higher than the *EPA*'s maximum safe limit of one (1) mcg/m^3 (for the residence). Even 30 minutes after a filling is removed, there can be levels of mercury vapor above *OSHA*'s maximum in the vicinity of the tooth being treated. You, the patient, could breath this vapor into your lungs where it could be absorbed by the blood stream and taken throughout the body. Also, while the dentist drills the tooth, small particles of silver mercury filling material scatters in all directions. You could absorb some of this mercury through the gums and other tissues of the mouth, and you can even swallow the tiny silver mercury particles. Many patients become very ill and their health symptoms amplify immediately after having silver mercury fillings removed. The additional high-level exposure is most likely the cause. Here are some of the most important precautions dentists can take to protect you from this additional exposure.

A. Have you breathe *oxygen* or *compressed air* through a nose piece rather than allowing you to breathe the newly released mercury vapor.

B. Isolate your teeth with a *rubber dam* and use a high-volume suction to keep the small mercury particles from going back into your mouth and throat and being swallowed.

C. Use *drapes* to cover your entire body, not just a bib that covers only your chest. Also, protective glasses are advised. These items help prevent small silver mercury particles from landing on your clothes and in your hair and eyes.

D. Provide as much *ventilation* as possible. Doors and windows can sometimes be left open, and a fan can be positioned to blow air across the patient's face while the dentist is removing the fillings. This helps dilute and remove the high concentration of mercury vapor and mercury mist being released close to the patient's breathing zone.

E. Provide for the administration of a slow intravenous drip of *Vitamin C* to bind (chelate) the mercury that inevitably gets through and finds its way into the blood stream.

Look for a dentist who acknowledges a concern for your health by doing everything possible to protect you from this additional mercury exposure. See Chapter 6 for more information on these physical protective barriers.

Another important area in interviewing a dentist is to check if he is caring for his own health.

11. *What precautions do you take for yourself (the dentist) to minimize your own exposure to the mercury vapor that is released when you remove silver mercury fillings?*

Dentists who are generally trying to help mercury toxic patients usually have symptoms of mercury toxicity themselves. This is mainly due to the fact that "a dentist is practically taking a bath in mercury everyday."

Even if he is not placing new silver mercury fillings but only removing old ones, the dental office may still be contaminated. High levels of mercury vapor are released every time the dentist removes a silver mercury filling. Microscopic particles of mercury scatter all over the dental office and land in the carpets and cracks of the floor and walls. It is practically impossible to completely remove mercury once it contaminates a treatment room; even if it could be thoroughly cleaned, it would be recontaminated every time another filling would be removed. The dentist essentially breathes the same air as the patient, and his face and body are being flooded with mercury all day long. Here are some of the precautions that the dentist should be taking for self-protection:

A. Wearing a special mercury filter mask while removing the fillings. This mask is not a thin blue surgical mask, but rather a thick, brown industrial mask that has been impregnated with charcoal.

B. Ventilation with a properly positioned fan to diffuse the mercury vapor being released.

C. Occasional IV Vitamin C sessions for the dentist.

D. Linoleum in the treatment room instead of carpet, for easier clean-up of mercury particles. (This is difficult to find since many dentists' offices were built before they recognized the problem.)

E. At least one separate set of clothing for treatment other than that which is worn to and from the office. Ideally, a dentist could wear a surgical gown or a white smock while removing silver mercury fillings. The mercury that gets into his clothes can be easily absorbed into his skin.

F. The dentist's vacuum pump (the actual 1-2 HP motor) is best if vented outdoors and not indoors. Otherwise, some of the mercury vapor that the dental assistant suctions can be recirculated into the air in the office. Again, this is difficult to find for the dentist may not have been aware of this at the time his office was designed. Unfortunately, many dental offices have their vacuum unit exhaust in a closet or under a counter inside the office suite. Ideally, both the vacuum pump and the compressor units should be outdoors as far away from the treatment room as possible.

G. Protective glasses and rubber gloves for the dentist and dental assistant are of extreme importance.

A word of warning: Forget any dentist who does not wear gloves!

That is for your protection as well as his. Don't believe any excuses here. There is almost no reason for not wearing gloves: in the extremely rare case when the patient (or the dentist) is allergic to rubber, the dentist can wear "vinyl gloves."

Look for the dentist who has realized his "Catch-22" situation.

Namely, the more he removes silver mercury fillings to detoxify his patients, the more he needs to detoxify himself because of the new exposure to mercury vapor that is released with his drill.

Otherwise, he may fall victim to mercury poisoning which can effect his abilities.

12. *How do you determine what dental materials to use for replacement of silver mercury fillings?*

In choosing a replacement filling material (porcelain, gold or plastic) the dentist must consider several factors: patient preference, cost, strength, flexibility and individual sensitivities. You may have read in Chapter 4 that solid porcelain is the best material. I believe that it is, but there are certain situations where it can't be used. Consider yourself fortunate if solid porcelain can be used for every filling in your mouth. Otherwise, you may need to be tested for sensitivity to dental materials. There are several ways to do this. Your dentist may work closely with an allergist or a clinical ecologist and refer you for what is called "Patch Testing." The drawback here is that you may become sensitized to a particular dental material simply from the testing itself. Accupressure can be used in some cases as well as kinesiology. Biocompatibility testing may become the wave of the future for dental material sensitivity evaluation.

13. Do you wear magnifying glasses (called loupes) and use lights (called fiberoptics) on your drills while removing silver mercury fillings?

Since dentists work in a small, dark, wet area of a tooth, it is easy for them to overlook small pieces of silver mercury filling material if they do not have proper magnification and lighting. Ask your dentist his feelings on the use of these for your treatment.

14. Do you have any satisfied patients who might be willing to talk with me about their experiences with mercury toxicity?

I can already see the letters that will be written to me for suggesting that you ask this question. There will be much resistance to the sacred doctor/patient privileges. However, there are patients who want and need to tell others about how they got better. Many patients volunteer to share their experiences with others once they have gratefully recovered. Besides, if you were going into the woods to fight a bear, wouldn't you want to talk with someone who had already been there and fought that bear? It won't hurt to ask if the dentist has any satisfied patients who have given him written permission to give out their names to others. Also, the dentist may be able to refer you to one of the "mercury toxic patient support groups" that have recently been formed.

15. Do you have silver mercury fillings in your mouth? What about your family?

It never ceases to amaze me when I discover dentists who address mercury toxicity while still having silver mercury fillings in their own mouths, not to mention the mouths of their loved ones. I confess, that I still had some in my mouth, too, when I first started seeing mercury toxic patients. However, as soon as I saw two or three successful cases, I rushed out to have myself detoxified of mercury. A dentist either completely believes in this issue or not. There is no sitting on the fence. You would not take smoking physicians very seriously if they told you smoking can contribute to lung cancer and that they could help you stop smoking, would you?

When it comes to mercury toxicity, don't take dentists seriously if they (or their family) are walking around with silver mercury fillings in their mouths.

Below I have listed all 15 questions for easy reference.

Questions for Interviewing Dentists

1. How long have you been opposed to the use of mercury in dentistry?

2. How did you get interested in mercury?

3. About how many mercury toxic patients have you helped by removing their silver mercury fillings?

4. How did you get your knowledge concerning mercury toxicity?

5. How do you keep your knowledge current concerning mercury toxicity?

6. What things do you do to test and monitor for mercury intoxication?

7. In what ways do you help a mercury toxic patient to detoxify the mercury that has been absorbed into the tissues of his body?

8. What other health care providers do you work with in relation to helping people with mercury toxicity?

9. Do you remove fillings in a particular order?

10. What precautions would you take to protect me of further mercury exposure from vapor and particles released while removing my fillings?

11. What precautions do you take for yourself (the dentist) to minimize your own exposure to the mercury vapor that is released when you remove the silver mercury fillings?

12. How do you determine what dental materials to use for replacement of silver mercury fillings?

13. Do you wear magnifying glasses (loupes) and use lights (fiberoptics) on your drills while removing silver mercury fillings?

14. Do you have any satisfied patients who might be willing to talk with me about their experiences with mercury toxicity?

15. Do you have silver mercury fillings in your mouth? What about your family?

CHAPTER 9

Deciding Which Dentist To Choose

N ow that you have the proper background information to determine the best dentist for you, it is time to begin interviewing. Continue your search until you find a dentist who can answer all your questions to your satisfaction. Any dentist who has been interviewed in this manner will know that you expect the very best possible care. Don't be surprised if a dentist refuses to accept you as a patient after your interview. The dentist may think you are too demanding or that your case is too complicated. That is fine. He has as much right to refuse you as his patient as you do in refusing to accept him as your dentist. If that should happen, don't take it too personally. You may have simply made him much more aware of patient care in the area of mercury toxicity.

Only you can make the final decision as to which dentist to entrust for your dental treatment and mercury detoxification procedures. Furthermore, only you can decide what compromises you can allow. However, if you have followed the suggestions on interviewing and have thoroughly read Chapters 6, 7, and 8 of this book, you should be armed with more than enough information to make an intelligent choice. Please do not let an impressive looking office or its proximity to your home make the decision for you. Remember, not all dentists are created equal. Hopefully, you have interviewed enough dentists that you have several at the top of your list. Pick one (or preferably two or three) that suits you the most and make another appointment for examination and testing.

Usually the dentist will ask you to make a separate appointment after your examination and testing are completed. This is when you will discuss the treatment plan and your investment. Of course, you don't have to accept the treatment plan. You may ask the dentist for copies of your laboratory test results if he hasn't already given them to you, and take them to the other dentists on your list. I recommend that you compare treatment plans with several dentists. Don't be surprised if they are all different. There is usually more than one way to design a particular dental treatment, and each has its pros and cons. Rarely, is there only one correct way to do anything, including practicing dentistry.

When you have finally chosen a dentist, give him your complete support. Let him know you have screened several dentists and that you believe he is the best for you. Tell him he has your confidence and that you trust him to do everything in his power to help you through this experience. Thank him for accepting you as a patient, because you and he both know you are going to be expecting more than the average patient. When he sees your attitude, he will give you nothing less than his best effort because he doesn't want to lose your confidence. He appreciates someone who recognizes his specialized talents.

If the dentist you finally choose measures up to your expectations, gives you 100% of everything he has to offer, and earns your confidence in recommending him to others, write him a letter telling him how you feel. He will greatly appreciate your letter.

One last thing: If you do write a letter to your dentist as suggested above, please send me a copy. I want to know where those "rare birds" are and what they are doing to be so special. (See the last page of Chapter 12 for Dr. Taylor's forwarding address.)

CHAPTER 10

The Patient's
Step-by-Step Guide to
Mercury Detoxification

1. Decide if your health symptoms correlate with the symptoms listed for mercury toxicity. Evaluate if testing would be of benefit to you. (See Chapter 1.)

2. Based on the information provided regarding the hazards of mercury released from your fillings, decide if you should consider preventive action. (Review Appendix A.)

3. Begin compiling a list of dentists and other health care providers who may be able to help you with mercury toxicity. Write to the organizations involved with this issue for names and addresses of their members. (See Chapter 7.)

4. Become very familiar with the specific aspects of mercury detoxification so you can evaluate any health care provider's level of expertise regarding therapy. (Study Chapter 6.)

5. Familiarize yourself with the testing procedures that are available for evaluating and monitoring mercury toxicity. (See Chapter 3.)

6. Learn as much as possible about the "Critical Link to Successful Therapy" namely nutrition and supplementation along with lifestyle factors. (See Chapter 5.)

7. Decide which dental materials would best suit your needs. (Review Chapter 4.)

8. Begin contacting dentists from your referral list regarding consultation appointments.

9. Compose a list of questions before your first appointment with the dentists. (See Chapter 8.)

10. Interview the dentists. Make notes about each dentist. Ask for copies of any x-rays or tests performed at your appointments.

11. If you are not able to find a dentist to suit all your needs, decide on what compromises you are willing to allow, if any, with regard to your dental care and detoxification therapies.

12. Make examination and treatment planning appointments with dentists that interest you the most. Ask for copies of all estimates and treatment suggestions.

13. Compare your notes and choose a dentist. (Refer to Chapter 9.)

14. Ask that blood, hair, and urine laboratory tests be performed if they were not already ordered by a dentist or other health care provider in Step 12 above.

15. Begin the nutrition counseling sessions well before any dentistry is performed.

16. Start supplementation and nutrition programs two to four weeks before any fillings are removed. In emergency situations a minimum of 48 hours on supplements, prior to treatment, is suggested.

17. Make appointments for the removal of your fillings. One or two appointments per week, (2 - 4 hours each), is ideal for most people. However, some sensitive patients must proceed much more slowly.

18. Three weeks after your fillings have been removed, it is imperative that follow-up blood and urine tests be taken to monitor your progress. Your health care provider will evaluate the results and advise you if any modifications for detoxification are necessary.

19. Have a three month follow-up testing performed and evaluated.

20. For the severely ill patients, six month re-evaluation appointments, with complete testing, is recommended.

21. Write a letter to your dentist about your experience. Please send a copy of your letter to me, Dr. Taylor. (See forwarding address at the end of Chapter 12.)

CHAPTER 11

Questions and Answers

D*o you recommend that everyone have their silver mercury fillings removed?*

This is a tricky question, and difficult to answer with complete honesty. Asking this question is similar to asking a cancer researcher or surgeon if he thinks everyone should stop smoking.

In theory, at least, if everyone had their fillings removed, there would be an overall increase in health for the population. However, this is not practical for many reasons, including the fact that there aren't enough dentists to undertake such a chore. Even if every dentist in the world stopped whatever he was doing and began to devote full time to removing people's silver mercury fillings, the job could not be done in one lifetime. There are just too many people with too many fillings, and the removal procedure is too slow for that to happen.

My usual answer to this question is, "No, not everyone should have their silver mercury fillings removed." But basically any person might consider having his silver mercury fillings removed especially if there are biological parameters that coincide with symptoms that indicate there would be a possibility for health improvement with removal.

In any case: don't have any more mercury placed in your mouth!

How can I be certain that the mercury in my fillings is harmful to my health?

A better question would be "how can I be sure that the mercury in my fillings is *safe*?" Direct that rephrased question to any dentist who is placing silver mercury fillings, or to any pro-mercury dental authority. Since mercury is a known poison, the burden of proof of safety should be on the shoulders of anyone who is advocating its use.

How can I know for sure if I am intoxicated with mercury?

Diagnosing mercury toxicity is difficult due to its insidious nature and the vagueness of symptoms. Some people can correlate the beginning of certain symptoms with the placement of silver mercury fillings. However, usually the symptoms of mercury toxicity appear gradually several months or even years after the initial exposure to mercury. Do you have classic symptoms of mercury poisoning? Having a body temperature below normal is the most consistent sign of mercury intoxication. Chronic fatigue is the symptom most often indicated by those having had mercury poisoning, followed by central nervous system disorders. Do you have unexplained emotional problems like depression, sudden anger, or irritability? Have you seen other health care providers and tried other therapies that have not satisfactorily helped your symptoms? Have you seriously tried to adopt a healthy lifestyle (no sugar, alcohol, cigarettes, or caffeine; plenty of sleep; regular physical activity; healthy diet with plenty of vegetables, fruits, grains, beans, seeds, and nuts; vitamin and mineral supplements) without any improvement of your symptoms? Do you have several large silver fillings in your mouth? Do you also have gold crowns? Do you have a metallic taste in your mouth? Answering "yes" to these and other related questions increases your chance of being mercury intoxicated. A qualified health care provider can test you and help you identify other indicators. The more indicators that are uncovered, the greater the possibility of mercury toxicity. You may not realize it, but your health care provider relies very heavily on what you tell him about your symptoms and about yourself to diagnose almost anything.

How fast does the mercury come out of the silver mercury fillings?

The average silver mercury filling contains 50% mercury. One dental researcher removed fillings that had been in a patient's mouth for 5 years. He also removed 20 year old fillings from another patient's mouth. He tested for percentage of mercury content on the surface of the fillings which included both the outside and the inside of the fillings. (The outside surface of fillings are more prone to vaporizing, thus being inhaled and taken into the lungs, or picked up by nerve endings in the nasal cavity. The inner portions of fillings are more prone to releasing mercury which will be absorbed like a sponge by the roots of the teeth, as well as the surrounding bone and adjacent gum tissue.) The 5-year old fillings showed they had lost one half of their mercury content from the surface, and the 20-year old fillings had lost all of their mercury. Another researcher found that, on the average, silver mercury fillings are replaced every eight and one half years. Fillings are usually replaced because they have fractured or the edges have begun to crumble away.

CHAPTER 12

One Final Word

I would like to leave my readers with one last comment concerning mercury's contribution to the symptomology of disease.

That is: do not expect the removal of your silver mercury fillings to be the answer to all of your health problems.

This therapy is not a total cure, but only one of many possible contributing factors in the process of disease.

We do get lucky sometimes. Every anti-mercury dentist has seen what seemed to be a miraculous "cure" immediately following the removal of a patient's silver mercury fillings. Sometimes it has happened after only one or two fillings are removed. These incidents are the exception, however, and not the rule.

Sometimes, a patient does not recognize significant improvements in symptomology for several weeks or even months to years. This is probably because the body needs time to remove the previously absorbed mercury and adjust physiologically. Of course, there are also cases when people have had only slight improvements.

The general public is becoming better educated every year and is doubting the traditional *single-cause/single-remedy* theory of disease. For example, the so-called "germ theory" of disease, that Louis Pasteur originated, is falling out of favor. There are, of course, situations where almost anyone would become sick from exposure to certain germs, as when eating food infected with botulism, but incidents like this are rare.

The average person is beginning to ask questions about disease in general. For example, why do some of the people working in an office "catch the latest flu bug" while some of their co-workers are symptom free? Why do some people get lung cancer after smoking for five or ten years while others have smoked for thirty or forty years and yet remain free of cancer? The answer to these questions lies in the fact that diseases are usually multi-factorial. People do not really "catch a flu germ." The "flu germ" catches them because their immune systems are depressed, and germs are opportunists. Why are some people's immune systems depressed? Mainly because of their lifestyles and environment. Poor nutrition, lack of sufficient sleep and physical activity, smoking, alcohol, drugs, caffeine, sugar, overeating, undereating, physical and emotional stress, polluted environment and poisons like mercury all take their toll on a person's immune system, as well as their organs and their bodies in general. Some people have strong constitutions and their bodies can take considerable abuse before health symptoms begin to surface. Others have more delicate bodies that respond negatively to almost any added stress. This is commonly referred to as a person's heredity.

If you happen to be one of those unusual persons who never seems to get sick and has no particular disease symptoms, you are one of the fortunate ones. Your body may be handling the chronic exposure to mercury from your fillings without difficulty (for the moment anyway). You may decide not to have your fillings removed. That is your right. Just remember that as you get older, your body's ability to deal with stresses may weaken.

If you are one of the growing population of prevention-minded people, you have enough information after reading this book to make an intelligent decision concerning your fillings.

On the other hand, if you are like the majority of people who are *"sick and tired of being sick and tired,"* you may be looking for some answers. Before you rush off and have your fillings removed, consider the other aspects of your life that could be contributing factors.

I am making this point because of the patients who do not get better. Most of them refuse to eliminate habits and lifestyle factors that are obvious major contributors to some of their symptoms. It's important to realize that, as previously discussed, there are other factors involved with symptoms of mercury toxicity besides silver mercury fillings. If you do decide to have your fillings removed, remember that each person is different. You have a different lifestyle and hereditary make-up than anyone else in the world. These factors will ultimately determine your response to silver mercury filling replacement. If you have your fillings replaced, at least you will have eliminated one source of potential danger to your health.

There is overwhelming evidence that mercury comes out of the fillings and gets into the body as well as unquestionable evidence of its toxic properties. The only thing missing is sufficient research on how little mercury it takes to cause damage. At this time, no one knows, and until we do know, mercury should not be used in dentistry.

Whether or not you decide to have your silver mercury fillings replaced, do not have any more mercury placed in your mouth or in the mouths of your loved ones!

I sincerely hope that this book will be of help to you in finding and choosing a dentist for detoxification of mercury. Your comments and suggestions will be greatly appreciated. Write to:

Joyal W. Taylor, D.D.S.
9974 Scripps Ranch Boulevard, Suite 36
San Diego, CA 92131

If you desire an appointment, call (619) 586-7626.

Mercury's Effect on Your Body

MERCURY AT THE CELLULAR LEVEL

A poison is any substance that impairs, injures, or kills an organism by its chemical action. All scientific authorities (scientists, toxicologists, metallurgists, immunologists, pathologists, doctors, etc.) agree that elemental metallic mercury (the type found in a common household thermometer), is a poison and can cause severe biochemical injury to living cells. When combined with certain kinds of compounds, however, mercury becomes even more poisonous. For example, an organic form of mercury called "methylmercury" is said to be 100 times more poisonous than elemental mercury.[1] This form of mercury is sometimes found in the algae-fish-human food chain and is one of the most potent and insidious poisons in existence.[2] Unfortunately, once elemental mercury has entered the body, various microorganisms that live naturally within the human body may be able to transform this elemental form into the even more deadly methylmercury.[3,4]

There is no known biological need for mercury in the human body, not even in trace amounts.[5]

Mercury accumulates in the body because once absorbed, it is unable to be excreted except in a very slow manner over extended periods of time.[6] No one yet knows how little mercury it takes to cause either temporary or permanent damage to human tissue.[7] What would you do if you learned there was a slow leak of a known poison like D.D.T., PCP, lead, carbon monoxide, etc., into your home? You would probably have the leak stopped or move to another house. Unless your reasoning ability had been altered, you probably would not spend very much time debating whether or not the particular form or dosage of that poison was going to damage your body today, next week, next year, or twenty years from now. You would simply want to eliminate that poison from your environment. Wouldn't you? In the same way, any prudent person who knows the facts about mercury's toxic properties would not want any amount of exposure in any form of this awful poison.

The purpose of this chapter is to explain what mercury actually does to make it so toxic to the body. To do this we must look briefly at its chemical effects. There are many avenues in the human body where mercury can damage individual cells. However, only a few basic ones will be discussed.

Sulfhydryl Bonds of Proteins

The most talked about area of concern is the attraction between sulfur and mercury.[8] Sulfur atoms (S) are found throughout the body as parts of large molecules in a form called sulfhydryl bonds (*SH*). Mercury (*H_g*) is greatly attracted to these bonds.[9,10] A mercury atom will remove a hydrogen (*H*) atom from the sulfhydryl bond and place itself into the site previously occupied by the hydrogen.[11] The simplified chemical formula for this reaction looks like this:

$$XS\text{-}H + H_g = XS\text{-}H_g + H$$

This reaction takes place very easily whenever mercury and sulfhydryl bonds come together. The resulting sulfur mercury bond (S-*H_g*) is strong and irreversible. The X in the previous equation stands for the rest of the large molecule to which the sulfhydryl bond is attached.

Why are we so concerned about the fact that mercury attaches to sulfhydryl bonds? Because the resulting molecule, XS-*H_g*, can no longer function properly and is said to be inactivated or poisoned. Keep in mind that sulfhydryl bonds are found in every system and, virtually, every cell in the body. Sulfhydryl bonds are predominately found in proteins throughout the body.[12] There are various combinations of 22 recognized amino acids that make up the protein molecules in humans. Three of the amino acids, cysteine, cystine, and methionine contain sulfhydryl bonds. Whenever a protein molecule containing any of these three amino acids comes into contact with mercury, the entire protein molecule is rendered inactive. For example, the cell membranes (cell walls) of all cells in the body are composed, in part, of protein molecules. What do you think happens to a cell wall when its protein is contaminated with mercury? Its function is altered. What is the function of the cell membrane? One function is to allow nutrients into the cell and waste products out of the cell. Another function of the cell membrane is involved with the exchange of oxygen and carbon dioxide. Therefore, when a cell wall becomes contaminated with mercury, the cell undergoes varying degrees of starvation, drowning and suffocating in its own waste products. Following, we will mention a few of the most significant protein molecules found throughout the body that contain sulfhydryl bonds and, consequently, can be affected by mercury.

Insulin is a large protein molecule that contains three sulfhydryl bonds.[13] It is produced in the pancreas and regulates the glucose (sugar) levels in the blood. The more insulin produced, the less sugar in the blood, and, conversely, the less insulin produced, the more sugar in the blood. The importance here is to recognize that mercury can attach to the sulfhydryl bonds of insulin, rendering it inactive and thus contributing to sugar imbalances.

Enzymes, as well as coenzymes, are proteins also containing sulfhydryl bonds.[14] Enzyme molecules are referred to as catalysts, which means they have the ability to initiate chemical reactions within the body. They may not be very large, but without enzymes, the body's energy producing reactions would not occur. Think of an enzyme molecule as you would the key to your automobile ignition. Without the key, you can't get the motor started. Until the motor starts, there is no gasoline combustion to produce energy and make the automobile move. Without enzymes to start energy-producing chemical reactions, there would be no biological cellular functions at all, and therefore, no life.

One vitally important enzyme which requires sulfhydryl bonds is acetyl coenzyme A. This enzyme is necessary for the metabolism of proteins, carbo-hydrates, and fats.[15] These substances are chemically broken down until they eventually find their way into a complex series of reactions called the Krebs Cycle. This cycle is recognized as the final stages of the breakdown of food and the actual production of energy. Just before foods enter the Krebs Cycle, they are broken down and converted to acetyl coenzyme A. When mercury is present, the sulfhydryl bond of *acetyl coenzyme A* may be inactivated and, thus, energy production inhibited. This may be one explanation for the most common symptoms of chronic, low grade mercury intoxication, namely tired-ness, lack of energy, oversleeping, inability to concentrate, and, in general, *chronic fatigue*. Acetyl coenzyme A is also necessary for the production of "good" cholesterol. Contrary to popular belief, not all cholesterol is bad. As a matter of fact, some cholesterol is absolutely essential to life. For example, hormones, bile, and even parts of nerve membranes are made from choleste-rol. Much of the brain itself is made from cholesterol. As you read about the nervous system, the endocrine system (glands and hormones), and the diges-tive system later in this book, keep in mind the fact that mercury can inhibit the production of good cholesterol.

Acetylcholine is also produced by an enzyme containing sulfhydryl bonds.[16] This substance must be present in the body in order for a nerve's electrical impulse to travel across a gap to another nerve ending. This gap is called a synapse.[17] The acetylcholine within the synapse is used up whenever an im-pulse travels from one nerve to another. New acetylcholine must then be made before another impulse can cross the synapse again. If its production is inhibited, symptoms of nerve dysfunction may result. The irregular heartbeat and muscle twitches that are often observed in mercury toxicity are an exam-ple of this symptomology.

There are many other protein molecules in the body that are involved with sulfhydryl bonds. However, enough information has been given for you to understand my point here. That is: mercury inactivates sulfhydryl bonds, thus it can interfere with the production and function of proteins which are found in every organ system and in every cell in the body.

Red blood cells contain hemoglobin which attracts oxygen and carries it to the cells. A hemoglobin molecule also has sulfhydryl bonds. Therefore, mercury can inhibit red blood cells from carrying oxygen. This is one explanation of how mercury can contribute to fatigue and anemia even though the red blood cell count may be in the normal range. Routine blood testing does not detect the mercury that may have incorporated itself into the hemoglobin by way of the sulfhydryl bonds.

Heredity

Heredity is passed from generation to generation by way of genes. Genes are complex molecules found inside the nucleus of every cell of the body. The genes are composed of double strands of deoxyribonucleic acid (DNA). These strands of DNA are composed of large molecules called nucleotides that are arranged in a specific pattern. This particular arrangement of nucleotides gives every cell its unique character. A brain cell and a heart cell may have essentially the same nucleotides that make up the DNA strands, the only difference being the particular arrangement of those nucleotides. There may be hundreds of nucleotides composing one DNA strand. When a cell duplicates itself, the nucleotides and their exact sequence within the DNA strand are replicated. The newly made DNA strands separate from the old strands and form the genes of the new cell, thus new life is formed and the genetic code is passed onto the next generation.

What happens if something causes the nucleotides to get out of sequence? The cell may be able to recover and repair itself if the disturbance is minor. If the sequence is broken at a critical area on the strand, that particular cell may die. At other times the cell may continue living, but not function as well as before. It may pass its altered nucleotide sequence to further generations as it duplicates itself. The aging process is sometimes explained in this way. The cell may even change into a cancer cell. So what does all of this have to do with mercury?

Mercury can cause single strand breaks and cross-linking of parts of strands of DNA.[18] For example, if the DNA of a newly developing embryo is affected in this way, the outcome may be miscarriage, stillbirth, or deformity. Mercury crosses the placental barrier and is actually found to be at higher concentrations in the fetus than in the mother.[19,20] Mercury is also highly attracted to brain and nerve cells,[21] which would explain the high incidence of neurological symptoms that are associated with mercury toxicity. For example, depression, irritability, and headache are common symptoms of mercury poisoning. Since brain and nerve cells generally do not duplicate themselves, the effect of mercury in the brain and nerves is more likely to cause permanent damage.

When mercury gets into the body, it becomes highly lipid (fat) soluble and forms lipid compounds with the cell membrane.[22,23] This allows the mercury to be quickly absorbed into the cell. Once inside the cell, mercury can modify DNA by interfering with the proper sequence of nucleotides.[24] Mercury's relation to the developing fetus as well as the brain and nerve tissue is discussed in more detail later in this chapter.

Other Effects of Mercury in the Body

Mercury is attracted to chloride and iodine. Therefore, it may affect the salt[25] balance and the thyroid functions.[26,27]
Mercury competes with magnesium for binding sites. If the body is low in magnesium, bacteria can produce twenty times more toxic substances of their own while in the body. Therefore, mercury intoxication may contribute to infections.

Calcium reacts to mercury. On the surface of cell membranes calcium contributes to cross-linking of proteins. Calcium also binds to mitochondria (the main energy producing organelle within cells). Therefore, mercury exposures may be a factor with symptoms related to decreased energy.

Selenium and mercury also compete for binding sites. Selenium is part of the protein, glutathione, which is an essential component of several enzymes. Insulin is made from glutathione, and mercury may inactivate insulin even though there is an adequate supply within the body.

Summary of Avenues of Mercury Damage Within the Body

The important issue to remember about these potential biological routes is that they provide pathways that permit mercury to travel all over our bodies upsetting very delicate, intricate lifegiving and sustaining functions in the process.

Don't believe that mercury in small doses can't hurt you.

WELL DOCUMENTED AND RECORDED HEALTH EFFECTS FROM CATASTROPHIC MERCURY OUTBREAKS

The literature is full of incidences of mercury contamination. For example, the miners in Almaden, Madrid the location of the richest mercury mine in the world) are known for their high incidence of early retirement because of the effects of the mercury vapor inhaled during their mining activities.[28] The miners take the dangers of mercury seriously and even have a special miner's hospital. Within the hospital is a special room that the miners call the "beach." The walls of the "beach" are lined with powerful heat lamps. Sometimes a miner will accidentally inhale mercury vapor while on the job and eventually develop tremors. Tremors are a sign that it is time to go to the "beach" for treatment of mercury intoxication. The patient strips and walks around and around the "beach" working up a good sweat. Some mercury is excreted through sweat glands and this treatment helps in the detoxification process. Some patients respond and return to work. Those who don't are pensioned. Some former workers cannot walk or feed themselves and a few shake so violently that they must be strapped into bed.

During the 19th century, workers in the felt hat industry dipped furs into vats of mercury solution to soften them for shaping. In the process, they not only inhaled the mercury vapor but they also absorbed mercury through their skin while handling the furs. The result was tremors, loosening and loss of teeth, difficulty in walking, mental disability, and incoherent speech. Legend has it the *Mad Hatter* in Lewis Carrol's *Alice's Adventures in Wonderland* was patterned after such a victim.[29]

Mercury poisoning has struck others in industry. Munitions workers, thermometer fillers, and laboratory technicians are just a few examples.

It was in 1953 that man was abruptly introduced to the deadly effects of mercury on a large scale basis. That year in Minamata, a small fishing village in Japan, the residents began to fall victim to a mysterious and catastrophic disease. Within three years, 30 patients had died from an unknown epidemic, and dozens more were seriously injured. At first local doctors made various diagnoses that this mysterious disease seemed to mimic: brain tumors, cerebral palsy, syphilis, and Japanese encephalitis. Most of the patients had become ill after eating fish. Some villagers reported seeing cats walking in circles and crows falling from perches. Both ate fish. Before the pathologist discovered the cause of this epidemic, 46 people had died and hundreds more were permanently disabled. What was the cause? Fish contaminated with *mercury*.[30]

How did the fish become contaminated? A large company in town was dumping mercury into the bay as a by-product in the manufacture of plastics such as vinyl chloride. Microorganisms were absorbing the mercury. Small fish were eating the microorganisms. Large fish were eating the small fish. And the villagers, as well as the animals, were eating the larger fish. The farther up the food chain, the more concentrated the mercury. Thus, people, cats, and birds eating the larger fish were affected the most. Many of the children born to mothers who had eaten the contaminated fish were born deformed and mentally retarded. Note that the mothers of these deformed babies generally did not become ill themselves. This was a clear indication that mercury crossed the placental barrier and, in fact, was more concentrated in the fetus than in the mother.

In other words, the mother may be protected from the effects of mercury by passing it to her developing baby.

Autopsies were performed on some of the animal and human victims. The findings in each case were the same — damage to neurons, resulting in fewer brain cells. These conclusions explained the symptoms perfectly; stumbling, convulsions, numbness of arms and legs, visual problems, difficulty in speaking, lowered intelligence, and mental retardation.

Japan experienced a second outbreak of mercury poisoning in 1964 at Niigata. There were six deaths out of a total of 47 cases identified. This second tragedy spurred the government to set up an agency to regulate industrial pollution, and no further incidences have occurred in Japan.[31]

Sweden also experienced the effects of mercury in the 1950's when various birds began to die from unknown causes. It seems that the farmers were treating their grain seed with mercury as protection against bacteria and fungus. The birds would eat some of the seeds during the planting season and, thus, they fell victim to mercury poisoning. In 1966 Sweden placed severe restrictions on the use of mercury in agriculture.

American scientists observed Japanese and Swedish experiences, and in 1969, the *Food and Drug Administration (FDA)* established guidelines for maximum limits of mercury found in fish. However, in 1970, mercury poisoning became a matter of public concern in the United States.

A farmer in New Mexico fed mercury-treated grain seed to his hogs.[32] Later he butchered one of the hogs and he and his family ate the meat. As a result, three of his children were severely crippled. His wife was pregnant at the time and the baby was born blind and retarded. During the same year high levels of mercury were reported in fish caught in the Great Lakes between the United States and Canada. Investigation showed that the dumping of waste products from plastic manufacturers was the source of the mercury contamination. This prompted fishing restrictions in the area. Also in the same year, canned tuna and swordfish were found to have high levels of mercury. The *FDA* ordered the withdrawal of 12 1/2 million cans of tuna and advised Americans not to eat swordfish. The *Department of the Interior* filed suit against operators of nine plastic manufacturing plants for polluting the waters with mercury. In time, more than 30 states reported mercury contamination in rivers and streams throughout the United States.[33]

Iraq had three major outbreaks of mercury poisoning involving the consumption of seed grain that had been treated with mercury. In 1956 approximately 100 people were hospitalized. In 1960 over 1,000 cases were reported, but by far, the worst tragedy ever to be reported occurred in Iraq in 1972.[34] There were over 6,350 persons hospitalized of which 459 died. Their deaths were directly attributed to eating mercury-contaminated seed grain that had mainly been consumed in the form of bread.

With the Iraqi and Japanese tragedies, it was the initial symptoms that assisted authorities in identifying mercury as the toxic agent involved. The effects of mercury were subtle at first but continued in severity for several months. Some symptoms did not surface for several months or more. Since the nervous system was the first to be affected, the intoxication usually produced varying degrees of the following symptoms:

1. Numbness of arms and legs

2. Tingling of toes, lips, nose

3. Fine tremors of the hands

4. Loss of ability to perform fine movement of hands

5. Failure of muscle coordination

6. Narrowing of the field of vision

7. Hearing impairment to complete deafness

The above are but a few of the most easily identified symptoms of mercury poisoning. When there is a slow continuous (chronic) exposure to mercury, there are dozens of other symptoms that begin to surface as the mercury continues its destruction on the brain, nervous system and practically all organs and systems throughout the body.

ORGANS AND BODY SYSTEMS AFFECTED BY MERCURY

Mercury does not affect all people in the same way. Each of us has our genetic strong points and our weak links. Mercury seems to search out the individual weak links and cause symptomology in those areas. Therefore, we can't always compare the symptoms of one individual with the symptoms of another. There may be some similarities of mercury toxicity between individuals, but usually the symptoms are varied from person to person. For example, we know that the central nervous system is usually the first area to show damage from mercury exposure. However, one person may get hand tremors while another person has hearing loss. Some people have a very low tolerance to mercury, while others show little or no signs of toxicity to small exposures. The following is a list of the most commonly affected tissues and organs in the event of mercury poisoning.

The Brain and Cranial Nerves

The brain is surrounded by a thin membrane called the "blood-brain barrier." This membrane allows an exchange of nutrients between blood and brain tissue. It also filters out many poisonous substances that are picked up by the blood from our food, water, and environment, as well as those toxic substances produced in our own bodies. Without this barrier, the brain would be totally vulnerable to thousands of poisons, and we could not live in our environment. Unfortunately, mercury can penetrate the blood-brain barrier, enter the cells of the brain from blood, and alter the barrier in such a way that other poisons can more easily penetrate the brain.

There is also another route to the brain tissue that mercury may use to bypass the blood-brain barrier. When mercury vapor is inhaled, the very fine nerve endings of smell receptors from the olfactory lobes can pick up the molecules of mercury. Since mercury has the ability to travel through a nerve, it has direct access to other parts of the brain by way of these olfactory lobes without ever having to encounter the blood-brain barrier.[35] The same principle is evident with the immediate action of "snorting" cocaine, airplane glue, or gasoline fumes. Mercury may travel more slowly through the nerve endings of the nasal cavity into the olfactory bulbs and then into various parts of the brain, than say, cocaine, but once it gets there it is difficult to remove.

Recently, it has been discovered by autopsy that persons with silver mercury fillings have three to four times as much mercury in the brain tissue as persons without silver mercury fillings.[36] Chapter 2 discusses how the mercury gets out of the fillings and into the brain and other organs.

Dentists who place silver mercury fillings are exposed to high doses of mercury vapor every day.[37] The vapor goes directly into their face and nose every time they place or remove silver mercury fillings. We did not realize how significant this was until recently when autopsies were performed on several dentists in Sweden.[38] Extensions of the brain called pituitary glands, were tested and found to have 300-400 times as much mercury in them as compared to those in non-dentists who had never had silver mercury fillings. This study cannot easily be reproduced for further evaluation, especially in the United States, because the pituitary is a highly prized gland and during autopsies it is normally saved for use in the pharmaceutical industry.

Kidneys

The kidney is another organ that suffers damage from mercury.[39,40] Autopsies show higher levels of mercury in the kidneys of persons with silver mercury fillings as compared to those of persons who do not have the fillings.[41] Because the primary function of the kidneys is to filter poisonous substances and by-products from the blood, they are exposed to higher concentrations of mercury than most other organs. While the kidneys are trying to protect the rest of the body from mercury by filtering it from the blood, they inadvertently absorb much of the mercury into their own cells. This may damage the kidneys and hamper their function.

The kidneys also remove some water from the blood collectively resulting in urine. Chronic exposures to mercury can also cause the kidneys to filter too much water (diuresis). At one time mercury was even used by physicians as a diuretic (that is, until the awful side effects were recognized). Many mercury toxic patients have symptoms of frequent urination and complain that they can't sleep through an entire night without having to get out of bed several times to urinate. As damage progresses, excess urination decreases to a stage of insufficient urination. When too little urine is produced, the body retains fluids, called edema. Edema contributes to further accumulation of waste products within the body, and a person can actually become poisoned from within.

Depending on the severity of the damage to the kidneys, mercury may contribute to high or low levels of protein in the urine. Interference with protein metabolism increases susceptibility of the body to infections. There is also something else unique to kidneys that needs to be mentioned with regard to the effects of mercury. During the filtering process, many minerals such as sodium, potassium, phosphate, and other nutrients are, at first, filtered out of the blood along with by-products and poisons. Normally, these desirable substances are separated from the undesirable substances and quickly reabsorbed back into the blood before leaving the kidneys. Mercury may interfere with this reabsorption process, thus contributing to nutrient deficiencies.

Laboratory tests on animals show that even smaller doses of mercury than those commonly suggested as safe can cause kidney damage in animals even though the animals tested appeared outwardly healthy.

Heart, and Blood Vessels

Contraction of heart muscles as well as the muscle surrounding the large arteries is dependent upon the presence of calcium. Mercury can block the action of calcium within cells. With less calcium to assist in contractions, it follows that the heart and arteries may remain dilated and not constrict efficiently. This could contribute to low blood pressure.

When the body detects low blood pressure, it releases adrenalin which, in turn, creates an increase in blood pressure by stimulating the heart and arteries to constrict. Adrenalin also causes an increase in heart rate and blood output. Mercury may inhibit adrenalin and thus contribute to low blood pressure. It has been demonstrated that mercury poisoning can cause an abnormal EKG,[42] and many patients complain of chest pains that are of unknown origin.

Hormones and Glands

Hormones are substances produced by the glands: thyroid, adrenals, parathyroid, pituitary, pancreas, pineal body, and sex glands. They exert specific effects in the body on metabolism, growth, maturity, and essentially control and influence almost all body processes. The fact that hormones are basically composed of proteins is one reason why mercury can cause problems with glandular functions. The pituitary and thyroid glands are particularly attracted to mercury.[43,44,45,46] Autopsies of these glands have shown higher concentrations of mercury than in kidney, liver, and even brain tissue.[47]

The pituitary gland is the master regulator of the other glands. It controls such things as growth, metabolism, reproduction, milk secretion, skin pigmentation, and water volume of urine. It also exerts a particular influence on the thyroid, adrenals and sex glands. Anything that can affect the activity of the pituitary gland has the potential to disrupt the delicate balance of practically all body processes.

The thyroid gland produces hormones that regulate energy production and metabolism. It also regulates tissue growth and development, as well as calcium levels in the blood. The thyroid uses iodine as an essential nutrient in the production and secretion of hormones. Mercury can interfere with the thyroid's ability to uptake iodine.[48,49,50] The sex gland hormones may be affected by mercury resulting in decreased sexual activity and infertility.[51]

Mercury may also impair the hormones secreted by the adrenal glands. Symptoms of adrenal dysfunction include intolerance to stress, anemia, low blood pressure, and diarrhea.

The pancreas produces insulin that regulates blood sugar. As discussed earlier in this chapter, mercury can inhibit sulfhydryl bonds such as those found in the molecules of insulin. Thus mercury may contribute to hypoglycemia and diabetes.

The parathyroid gland regulates mineral balance, and in particular, calcium, magnesium, and phosphate. Mercury is known to inhibit calcium and compete with magnesium for binding sites.

Depending on the severity of intoxication, mercury may increase or decrease hormone production of glands and thereby disrupt the intricate function that maintains the body's balance. This could explain the wide range of symptoms associated with mercury toxicity.

White Blood Cells, Allergy, and the Immune System

The white blood cells (lymphocytes) are responsible for both immunity and allergy. Immunity refers to the body's ability to fight off infection and disease. Allergy is present when the body's immune system overreacts to a particular stimulus.

The lymphocytes can be divided into what is called T-Cells (Thymus derived) and B-Cells (bone marrow derived). The T-Cells can be further subdivided into T-Helper Cells and T-Suppressor Cells.

When a foreign substance like a bacteria enters the body, T-Cells recognize the invader as a potential danger. They signal to the B-Cells which then produce protein substances called antibodies. The antibodies produced by the B-Cells are very specific to each new type of invader. The antibodies attach to the foreign substance. A T-Helper Cell then engulfs and destroys the invader. After all those particular invaders are destroyed, the T-Suppressor Cells signal to the T-Cells to discontinue sending messages to the B-Cells, and production of that particular antibody is stopped. The T-Suppressor Cells clean up most of the excess antibodies, but a small amount remains in the blood stream. If the same foreign substance should enter the body at some time in the future, the remaining antibodies will immediately attack before the invader can cause any problems. This is what is meant when we say that we are immune to something.

Sometimes the B-Cells are stimulated to produce a particular antibody that causes the release of histamine. This results in a red flushed skin reaction and is considered an allergic response. There is a delicate balance that must be maintained between T-Helper Cells, T-Suppressor Cells, B-Cells, and antibodies for optimum health. For example, if there are too many T-Suppressor Cells or too few T-Helper Cells, a person will become very susceptible to colds, the flu, and infectious diseases in general. As the imbalances get more serious, a person becomes susceptible to cancer, AIDS, Hodgkin's Disease, and a variety of other immune deficiency diseases.

Mercury may cause changes in the DNA (genes) of white blood cells.[52] This results in an alteration of cell function, impaired ability to duplicate itself, and even death of the cell.

Patients with silver mercury fillings sometimes have an imbalance of the ratio of T-Helper/T-Suppressor Cells. Upon removal of their fillings, a ratio approximating a more optimum state can oftentimes be achieved.[53]

Reproductive System

Mercury's effect on reproduction has been clearly recorded from studies of infants born to mothers who were pregnant during mercury catastrophes.[54,55,56,57,58] For example, after the Iraqi epidemic in 1972, of the 32 infants born to mothers who had eaten the mercury contaminated grains, approximately 32% of the babies were born with cerebral palsy, and 28% died during infancy. Over one-half of the babies who survived had delayed mental and motor development and one-fourth of the survivors had abnormally small heads. Another example is the high incidence of abnormalities that occurred in the children born to victims of the mercury crisis of Minamata, Japan, in 1953. Abnormalities such as brain damage, arm and leg deformities, disturbances of body growth, speech difficulties and mental retardation, occurred in the infants of mothers who had eaten the mercury contaminated fish during pregnancy. Mercury can injure the nervous system of an unborn child even at levels which are considered safe for adults.

There are higher levels of mercury in the blood of a fetus than there are in the mother's blood. This means that during pregnancy, the fetus absorbs more than an even share of mercury from the mother. Mercury not only crosses the placental barrier, but is excreted through the mother's milk.

The medical profession recognizes the fact that exposure to mercury during pregnancy is potentially hazardous, especially to the unborn child. Obstetricians are now cautioned to advise their patients to avoid all exposure to mercury. It is hoped that this book will help both doctors and patients recognize that, at least during pregnancy, the placement of new silver mercury fillings should be avoided.

Nervous System

There are virtually no tests available to recognize preclinical disease states of the central nervous system. It is known that mercury damages brain and nervous tissue, but reports of the mental and emotional problems that result are obtained through medical histories, complaints of the victims themselves, and observations made by doctors and researchers. Some of the symptoms related to the nervous system that have been recorded from the victims of mercury epidemics include headache, depression, irritability, dizziness, and sleep disturbances.[59,60,61] (See Chapter 1 for a more complete list of mental and emotional symptoms that have been related to mercury toxicity.)

Patients and doctors are encouraged to recognize mercury intoxication as a potential causative or contributing factor in the diagnosis and treatment of these symptoms.

One final note in considering mental and emotional problems concerns the electric current generated by silver mercury fillings.[62] Whenever two or more metals are placed in a liquid (an electrolyte), a battery is essentially formed. Since silver mercury fillings contain 5 or more metals, when combined with a person's saliva (which acts as an electrolyte), a type of battery is also created. Voltage and amperage meters have been used to measure the amount of current being generated by the silver fillings. Tests show that just one silver mercury filling can generate many times the amount of electrical energy that the brain generates during its normal functioning. Anyone with silver mercury fillings who has accidentally chewed a piece of aluminum foil from a gum wrapper, has experienced this electrical phenomenon. No one knows for sure the impact of this electrical current that is being produced by silver mercury fillings, but since the teeth are very close to the brain, many people believe that fillings are generating a chronic low-grade "electrocution" of the brain and nervous system. This is one explanation for the relief of headaches, sleep disturbances, nightmares, irritability, etc., that some people experience after the proper removal of their silver mercury fillings.

The Mouth

Mercury intoxication can cause a metallic taste in the mouth and an increased flow of saliva. The salivary glands and tongue can also become enlarged. Mercury has been shown to be a contributing factor in gum disease, bad breath, bleeding gums, mouth ulcers, loosening of teeth, and allergic tissue responses in the mouth.[63,64,65]

When biopsies have been performed on roots of teeth containing silver mercury fillings, extremely high levels of mercury have been present.[66,67] Bone and gum tissue surrounding teeth with silver mercury fillings have also been biopsied with similar findings.[68,69] Purplish-black pigmentation areas called "amalgam tatoos" (argyrosis) appear in the gum tissue immediately surrounding some teeth with large silver mercury fillings. In some instances, amalgam tatoos may appear on cheek and lip areas.

References

[1] Ely, T., "Methyl Mercury Poisoning in Fish and Human Beings," *Modern Medicine*, November 16, 1970, pp. 135-141.

[2] Ibid.

[3] Heintze, V., et al, "Methylation of Mercury From Dental Amalgam and Mercuric Choloride By Oral Streptococci in Vitro," *Scandinavia Jouranl of Dental Research*, 1983, 91 (2): 150-152.

[4] Rowland, I., et al, "The Methlation of Mer4curic Choloride By Human Intestinal Bacteria," *Experientia*, 1975, 31:1064-1065.

[5] Kirschmann, J., Nutrition Almanac, 2nd Edition, New York: McGraw Hill, 1984.

[6] Sugita, "The Biological Half-Time of Heavy Metals. The Existence of a Third 'Slowest' Component," *International Archives of Occupational Health*, 1978, 41:25-40.

[7] Battistone, G. et al, "Mercury: Its Relation to the Dentist's Health and Dental Practice Characteristics," *Journal of the American Dental Association*, 1976, 92:1182.

[8] Rupp, N. et al, "Significance to Health of Mercury Used in Dental Practice:

A Review," *Journal of the American Dental Association*, June 1971, 82 (6): 1401-1407.

[9] West, E., et al, Textbook of Biochemistry, New York: The Macmillan Co., 1957, p. 853.

[10] Vallee, B., et al, "Biochemical Effects of Mercury, Cadmium, and Lead," Annuals of Review Biochemistry, 1972, 44:91-128.

[11] Brown, W., Introduction to Organic Biochemistry, 2nd Edition, Boston: Willard Grant Press, 1978, p. 120.

[12] Rupp, N., op cit.

[13] Brown, W., Introduction to Organic Biochemistry, 2nd Edition, Boston: Willard Grant Press, 1978, p. 120.

[14] Rupp, N., op cit.

[15] Guthrie, H., Introductory Nutrition, 4th Edition, St. Louis: Mosby, 1979, pp. 341-347.

[16] Miyamoto, M., "Hg^{2+} Causes Neurotoxicity at an Intracellular Site Following Entry Through Na and Ca Channels," *Brain Research*, 1983, 267: 375-379.

[17] Komulainen, H., et al, "Effect of Hevy Metlas on Dopamine, Noradrenalin and Serotonin Uptake and Release in Rat Brain Synapotosomes," *Acta Pharm and Tox*, Volume 48, 1981, pp. 119-204.

[18] Verschave, L., et al, "Genetic Damage By Occupational Law Mercury Exposure," *Environmental Research*, Volume 12, 1976, pp. 306-316.

[19] Kuntz, W., et al, "Maternal and Cord Blood Background Mercury Levels: A Longitudinal Surveillance," *American Journal of Obstetrics and Gynecology*, Volume 143, Issue 4, 1982, pp. 440-443.

[20] Tejning, S., "Mercury Levels in Blood Corpuscles and in Plasma in 'Normal' Mothers and Their Newborn Children," Report 68 02Z, Department of Occupational Medicine, University Hospital, Lund, Sweden, Lung Stencils, 1968.

[21] Niosh, "Criteria for a Recommended Standard: Occupational Exposure to Inorganic Mercury," U.S. Department of Health, Education and Welfare, Public Health Service, 1973.

[22] Nakada, S., et al, "Susceptibility of Lipids to Mercury," *Journal of Applied Toxicology*, 1983, (3): 131-134.

[23] Magos, L., "Mercury — Blood Interaction and Mercury Uptake By the Brain After Vapor Exposure," *Environmental Research*, 1967, 1:323-37.

[24]Tiskesjo, G., "The Effect of Two Organic Mercury Compounds on Human Leukocytes in Vitro," *Hereditas*, Volume 64, 1970, pp. 142-146.

[25]Miyamoto, M., op cit.

[26]Suzuki, T., et al, "Affinity of Mercury to the Thyroid," *Industrial Health*, Volume 4, 1970, pp. 69-75.

[27]Trakhten, Berg I., Chronic Effects of Mercury on Organisms, Washington, D.C.: U.S. Department of Health Education and Welfare, Public Health Service, National Institues of Health, Superintendents of Documents, U.S. Government Printing Office, 1974.

[28]Putnam, J., "Quicksilver and Slow Death," *National Geographic*, October 1972, pp. 510, 512.

[29]Ibid., p.511.

[30]Ibid., pp. 514-516.

[31]Ibid., p. 517.

[32]Ibid., pp. 518-521.

[33]Ibid., p. 523.

[34]Damuji, S., et al, "Intoxication Due to Alkyl-Mercury Treated Seed — 1971-1972 - Outbreak in Iraq, Clinical Aspects," *Bulletin, World Health Organization*, Volume 53, (Suppl.) 1976, p. 65.

[35]Stortebecker, P., *Mercury Poisoning From Dental Amalgam Hazard to Human Brain*, Taby/Stockholm, Sweden: Startebecker Foundation For Research, 1985, S-18335.

[36]Schiele, R., et al, "Studies on the Mercury Content in Brain and Kidney Related to Number and Condition of Amalgam Fillings," Institution Occupational and Social Medicine, University Erlangen, Nurnberg, Symposium, March 12, 1984.

[37]Chou, N. et al, "Quantitative Determination on Mercury Released From Pure Mercury and Amalgam," American Dental Association Abstract #1282, IADR Abstracts, 1984, p. 313.

[38]Nylander, M., et al, "Mercury in Pituitary Glands of Dentists," *Lancet*, February 20, 1986.

[39]Nicholson, J., et al, "Cadmium and Mercury Nephrotoxicity," *Natune*, Volume 304, 1983, pp. 633-635.

[40]Stock, A., "Der Quecksilberge Halt des Menschlichen Organisms," *Biochemische Zeitschrift*, 1940, 3041:73.

[41]Schiele, R., et al, op cit.

[42]Dahhan, S. et al, "Electrocardiographic Changes in Mercury Poisoning," *American Journal of Cardiology*, Volume 1, 1964, pp. 178-183.

[43]Suzuki, T., et al, op cit.

[44]Trakhtenberg, I., op cit.

[45]Stock, A., op cit.

[46]Kosta, et al, "Correlation Between Selenium and Mercury in Man Following Exposure to Inorganic Mercury," *Natune* (London), Volume 254, 1979, pp. 238-239.

[47]Nylander, M., et al, op cit.

[48]Suzuki, T., et al, op cit.

[49]Trakhtenberg, I., op cit.

[50]Kista, et al, op cit.

[51]Burton, B., et al, "Acute and Chrnoic Methyl Mercury Poisoning Impairs Rat Adrenal and Testicular Function," *Journal of Toxicity and Environmental Health*, Volume 6, 1980, pp. 597-606.

[52]Verschave, L. et al, op cit.

[53]Eggleston, D., "Effect of Dental Amalgams and Nickel Alloys on T-lymphocytes: Preliminary Report," *Journal Pros Dent.*, 1984, 51:617-623.

[54]Kuntz, W., et al, op cit.

[55]Tejning, S., op cit.

[56]Amin-zaki and Clarkson, et al, "Prenatal Methylmercury Poisoning," *American Journal Disabled Children*, Volume 133, 1979, pp. 172-177.

[57]Tedeschi, L., "The Minamata Disease," *American Journal of Forensic Medicine and Pathology*, Volume 3, Issue 4, 1982, pp. 335-338.

[58]Rodier, P., et al, "Mitotic Arrest in the Developing Central Nervous System After Prenatal Exposure to Methylmercury," *Neurobehav Toxicol Teratol*, 1984, 6 (5): 379-385.

[59]Damuji, S., et al, op cit.

[60]Amin-zaki and Clarkson, et al, op cit.

[61]Tedeschi, L., op cit.

[52]Schriever, W., et al, "Electromotive Forces and Electrical Currents Caused by Metallic Dental Fillings," *Journal Dental Research*, 1952, 31:205.

[63]Goodman and Gillman, The Pharmacological Basis of Therapeutics, 6th Edition, New York: Macmillian Publishing Co., 1980.

[64]Shafer, W. et al, A Textbook of Oral Pathology, W.B. Sanders Co., 1974.

[65]Berhow, R., editor, The Merck Manual, 14th Edition, Merck and Co., Inc., 1982.

[66]Till, T., et al, "Mercury in Tooth Roots and in Jaw Bones," 1978, ZWR 87 (6):288-290.

[67]Soremark, R. et al, "Influence of Some Dental Restorations on the Concentrations of Inorganic Constituents of the Teeth," *Acta Odontal Scand*, 1962, 20:215-224.

[68]Freden, H. et al, "Mercury Content in Gingival Tissues Adjacent to Amalgam Fillings," *Ondont Revey*, 1974, 25 (2): 207-210.

[69]Till, T., et al, "Zum Nachweis Der Lyse Von Hg Aus Silberramalgam Von Zahnufullungen," *Der Praktische Arzt*, 1978, 32:1042.

The Controversial History of Mercury's Use in Dentistry

There is evidence that mercury was used in dental filling material by the Chinese as far back as 659 A.D.[1] Mercury in dentistry sprang out of a need for an inexpensive, easy-to-manipulate tooth filling material. Gold was the usual choice. However, only wealthy people could afford gold fillings and the handling of gold was difficult and time consuming.

In England, around 1819, dentists began to experiment with metal "pastes."[2] They used combinations of tin, bismuth, lead, and mercury. With what we know today about the toxic effects of lead, imagine allowing a dentist to deliberately introduce this poisonous material into your body. Well, mercury is far more toxic than lead![3]

The biggest problem with the first attempt at using metal "pastes" was that the metals had to be heated until they melted together, and then the mixture was "poured" into the tooth cavity. This, of course, damaged and even killed teeth (not to mention what may have happened if drops of the hot metal splashed onto the gums, cheeks, tongue, etc.).

Around 1826, a dentist in Paris began using a "silver paste" that did not require melting.[4] He would first grind silver coins, which also contained copper and other metals, into a fine powder. Since mercury remains liquid at room temperature, he could mix the powder with the mercury and obtain a paste-like material that could easily be pushed into a tooth cavity. The mixture would harden after a few minutes. This silver paste had problems, too, however. Upon setting, the mixture would expand, causing fractured teeth and/or bite interferences. Slight adjustments were made in the "silver paste" until the expansion problems were reduced. Later this mixture was called *silver filling material* or *amalgam*. These fillings were introduced into the United

States in 1833, and something of tremendous significance happened at that time.

It seems that two entrepreneurs from France, the Crawcour brothers, decided to set up a dental practice in New York City. These men were not licensed dentists, and they apparently had no formal dental training or experience in dentistry.[5] They were, what some people might call "*quacks*". (See Appendix C for an explanation of the origin of the word "*quack*".) They began using the new silver mercury fillings as a substitute for gold. The Crawcours were very successful, and their office attracted many patients, even some who could afford gold.

The established American dentists in New York City began to denounce the new silver mercury filling material. They knew that it was a poor material and that it had many drawbacks. These dentists attacked silver mercury fillings as a cause of *mercurial poisoning* as well as a cause of gum disease and other disastrous effects.

In 1834, these dentists formed a local organization called "*The Society of Dentists of the City and State of New York.*"[6] The organization was probably formed to combat the activities of the notorious Crawcours, but it experienced internal conflict over the mercury issue and had to be dissolved in 1839. The Society, however, had made such a fuss with the Crawcours that the brothers had to close their dental office only months after opening.

In 1840, organized dentistry in America formed "*The American Society of Dental Surgeons (ASDS).*"[7] The *ASDS* was so strongly opposed to the use of mercury that it required its members to sign "pledges" promising not to use it. In fact, in 1848, eleven members were suspended from the society in New York City with a claim of "malpractice for using amalgam." Nevertheless, some dentists, continued using the poisonous material.

The *ASDS* was plagued with in-fighting over this issue until 1855 when it withdrew most of its resolutions prohibiting the use of silver mercury fillings. This was a last ditch effort to save the dental society. It was too late, however, and the *American Society of Dental Surgeons* dissolved itself in 1856.[8]

Out of this strife arose a new dental organization,"*The American Dental Association,*" (the same *ADA* that is functioning today). This new organization wanted to unite all dentists, so it had no requirements for pledges against the use of mercury. Since the *ADA* did not oppose the use of mercury, dentists must have incorrectly assumed that silver mercury fillings were safe to use. As the years went by, more and more dentists began using silver mercury filling material until the majority of *ADA* members were using it.

Although the detrimental effects of mercury were well known in the 1800's, there was no inexpensive substitute for gold fillings except for the silver mercury fillings. Also, there was no governmental funding or interested scientists to conduct so-called "double-blind" studies to "prove" whether or not mercury in tooth fillings was safe. The use of mercury in dentistry was, however, quietly opposed by a few dentists and researchers.

In the late 1800's and early 1900's, a dentist by the name of Dr. G.V. Black was credited with the technical perfection of silver mercury fillings (among other things). He was, and still is, known as the "*father of scientific dentistry.*" Even today, it is almost considered "heresy" to oppose Dr. G.V. Black on anything to do with dentistry. Actually, Dr. Black did not concern himself with the possible toxic effects of materials placed into people's teeth. His particular expertise was in the meticulous design of cavities, the filling of those cavities, and the formulation of a mixture of metals that is considered the standard for silver mercury fillings today. He was a genius with technical and artistic talents. The fact that mercury is a highly toxic heavy metal capable of causing tremendous damage to the human body may never have occurred to him. Mercury in dentistry was not particularly a "hot" issue during Dr. G.V. Black's time.

Later in 1926, a highly regarded chemist from Germany, Dr. Alfred Stock, began conducting studies on the disastrous effects of mercury on the body.[9] He reported his personal experiences from mercury poisoning. He was poisoned partly by mercury vapors from mercury-containing instruments in his laboratory and partly from the release of mercury vapors from his dental fillings. Dr. Stock was the first to recognize that mercury vapor was constantly emitted from silver mercury fillings. He concluded that the mercury values he measured from the breath of patients with silver mercury fillings often surpassed the measured amounts from leaking, mercury-containing devices found in the laboratory. For many years Dr. Stock suffered from slowly developing mercury poisoning until he succeeded in diagnosing the cause of his symptoms which he summarized as follows:

First Stage   Predominantly psychological symptoms. Fatigue. Diminished working capacity. Irritability. Slight swelling of mucous membranes in the upper nasal region.

Second Stage   Extreme fatigue. Lack of concentration. Impaired memory for names, numbers, etc. Irritability and moodiness. Sensation of sheer "stupidity." Nasal obstruction with dryness, and a "stuffed nose." Nasal discharge (viscous, sticky, sometimes bloody). Ringing in the ears. Hearing impairment. Headache, often in front of head. Bleeding gums. Irregular heartbeat. Diarrhea. Slight tremor.

Third Stage   Headache. Dizziness. Vertigo. Tremor. Mental incapacity. Depression. Back pain. Diarrhea. Nasal discharge, sometimes bloody and sometimes crusty. Loss of smell. Bleeding gums. Loose teeth. Increased salivation.

Dr. Stock moved to a new laboratory that was free of mercury contamination and replaced all his silver mercury fillings. Only then was he able to recover from his severe mercury intoxication.

Dr. Stock published over 30 articles condemning the use of mercury in dentistry. His research methods were very accurate and his experiments are too numerous to mention. He gained academic and public support and probably would have single handedly helped dentists come to the long overdue conclusion that mercury does not belong in dentistry. However, World War II put an end to Dr. Stock's research and momentum. His laboratory and most of his records were destroyed during the war. The world was not interested in the effects of mercury in silver fillings with issues like world wide destruction and radiation poisoning with which to contend. Dr. Stock and his research died in near obscurity. However, a few dentists believed Dr. Stock and quietly continued "mercury-free" dentistry.

Since the early 1980's, more and more dentists, scientists, and other health care providers have been questioning the logic behind using a known poison, like mercury, in tooth fillings. The anti-mercury forces are growing so rapidly that I believe that a generation from now mercury will be banned from use in dentistry. The general public, as well as organized dentistry, will cringe to think that such an awful poison was ever allowed to be used in the human mouth in the first place.

Special Note: As this book is being written, Sweden's Government Health board has declared silver mercury amalgam to be "toxic and unsuitable as a dental filling material." It has recommended that the use of amalgam in dentistry be discontinued as soon as possible. As a first step, amalgam restorations "in pregnant women shall be stopped in order to prevent mercury damage to the fetus." With regards to previous statements concerning the safety of amalgam, the Division Chief, Viking Falk, has publicly stated, that "We now realize that we have made a mistake. This has caused people to suffer unnecessarily."

REFERENCES

[1] Ring, M., Dentistry: An Illustrated History, New York: Harry N. Abrams, Inc., 1985.

[2] McGehee, W. et al, A Textbook of Operative Dentistry, 4th Edition: McGraw-Hill, 1956.

[3] Sharma, R. et al, "Metals and Neurotoxic Effects: Cytotoxicity of Selected Metallic Compounds on Chick Ganglia Cultures," *Journal Comp Path*, 1981, 91:235-244, 1981.

[4] McGehee, W. et al, op cit.

[5] Ring, M., op cit.

[6] Ring, M., op cit.

[7] Ring, M., op cit.

[8] Ring, M., op cit.

[9] Stock, A., "The Hazards of Mercury Vapor," *Zeitschr Angew Chem*, 1926, 39:461-488.

Quacks and Mercury

M ercury is also called *quicksilver* and in some European countries it is spelled *quacksalver*. In the dictionary, a quack is defined as one who pretends to cure disease, and a salve is a substance for application to wounds or sores. The derogatory term *quack* may have first been used in reference to anyone using mercury preparations on the skin to supposedly "cure" diseases.

In the fifteenth to seventeenth centuries, mercury ointments and mercury baths were used as treatments for syphilis. These treatments seemed to temporarily help with the skin eruptions of syphilis, which, as we know today, were only the outward signs of the first phase of worse things to come. The mercury treatments caused disastrous side effects that were sometimes much worse than the disease they were trying to treat. This was a classic example of the infamous saying, *"The operation was successful, but the patient died."* The patients treated with mercury did sometimes die from the side effects, with painful and violent deaths.

Out of the false belief that mercury cured syphilis came the belief that mercury could cure other diseases for which there was no medical approach. The use of mercury increased dramatically in the 1800's. It was used as a treatment for just about anything from diarrhea to typhoid fever. It was discovered that mercury killed bacteria, and could be used as an antiseptic. Of course a poison as strong as mercury not only killed bacteria, but when applied to human flesh, killed living cells and sometimes killed the patient.

The next time you hear of someone who is using mercury, whether it be a dentist placing silver mercury fillings, or an advertisement for hemorrhoidal ointments which contain mercury, think about the origin of the word *quack*.

Non-dental Mercury Exposures Which May Be Affecting Your Health

A lthough your fillings are the main source of chronic exposure to mercury, they are not the only source. Most people in the civilized and industrialized countries are exposed to several other sources of mercury on a daily basis. Patients seeking therapy for mercury toxicity must not only have their fillings properly removed, they must also remove as many other sources of mercury from their environment as possible, especially during the initial treatment phase. Several months later, less seriously ill patients may find that they can safely withstand an occasional exposure to other sources; however, the most sensitive patients may be so sensitized that they will still have severe reactions to any mercury exposure for many years.

Mercury is used in medical as well as dental fields. It is sometimes used in drugs and over the counter medicines such as diuretics (water pills), antiseptics, hemorrhoidal ointments, skin ointments, laxatives, and contact lens solutions.

Check every ingredient on anything in your medicine cabinet for the words *thimerosal* or *sodium ethyl mercur*. These are mercury compounds used as preservatives and antiseptics. You can find thimerosal in dozens of products from eye drops to contraceptive gels. Optometrists are very aware of the problems experienced with thimerosal since many of their patients get eye irritations from contact lens solutions containing this compound. If you wear contacts, ask your optometrist to suggest "non-thimerosal" products. Sodium ethyl mercur is found in some "over the counter" antiseptics such as mercurochrome and methiolate. Psoriasis ointments and even toilet papers made from recycled paper may contain mercury.

Mercury can also be found in certain cosmetics such as mascara. It is used widely as a pigment in paints and dyes. Do you have tatoos? If so, you may have another chronic source of mercury exposure in your body in addition to your silver mercury fillings.[1]

Mercury may also be found in various products used in and around the house. Such things as fabric softeners, floor polishes, wood preservatives, adhesives, mercury cell batteries and fungicides may all contain mercury. Most people realize that some household thermometers contain mercury, but many are not aware that fluorescent lights contain even more.

Fish and sea foods of any kind may have high levels of the most dangerous compound of mercury, "methylmercury."[2] Of particular concern is tuna, swordfish, and all shellfish, including, shrimp, lobster, crab and oyster. Kelp and other seaweeds also have high mercury levels. For patients undergoing treatment for mercury toxicity, it is recommended that they refrain from eating any fish, seafood, or seafood products. The highly sensitive patients may have to eliminate fish and seafoods from their diet for a number of years.

Industry and agriculture use more mercury each year than all other sources combined. Mercury is included in everything from the manufacture of vinyl chloride to pesticides. Mercury poisoning has also become a problem because of environmental pollution. Each year there is more and more mercury in the air, water, and soil, mainly due to its use in agriculture and industry.

People are exposed to high levels of mercury vapor by simply walking into many dental offices[3,4] and should be particularly cautious of any dentist's office where new silver mercury fillings are being placed on a routine basis. Some of these offices will have mercury vapor levels exceeding the *Occupational Safety and Health Agency*'s maximum allowable limit.

The occupations listed below may also be sources of mercury exposure for many people.[5]

Barometer manufacturers
Battery makers
Dentists and dental personnel

Electroplaters
Embalmers
Farmers
Fur manufacturers
Gold and silver extractors
Ink producers
Jewelers
Makers of fluorescent and neon light bulbs
Manufacturers of explosive caps
Manufacturers of fireworks
Manufacturers of pressure gauges
Painters and paint manufacturers
Paper manufacturers
Photographers
Switch makers
Taxidermists

There may also be other sources of mercury in your environment which are difficult to identify. Not every source can be avoided, but during therapy for mercury toxicity, every effort should be made to eliminate as much exposure as possible.

As you can see, each person has different environmental sources of mercury exposure depending on their lifestyles, living quarters, eating habits, personal habits, occupation, etc. The health care provider who takes your health history and counsels you on mercury toxicity should allow sufficient time to help you identify the other sources that are unique to your particular situation.

REFERENCES

[1] D'Itri, R. et al, *Mercury Contamination: A Human Tragedy*, New York: John Wiley and Sons, 1977.

[2] Eyl, T., "Methyl Mercury Poisoning in Fish and Human Beings," *Modern Medicine*, November 16, 1970, pp. 135-141.

[3] Jones, D., et al, "Survey of Mercury Vapor in Dental Offices in Atlantic, Canada," *Canadian Dental Association Journal*, 1983, 49(6): 378-394.

[4] Chou, N. et al, "Quantitative Determination on Mercury Released From Pure Mercury and Amalgam," American Dental Association, Abstract #1282, p .313, IADR Abstracts, 1984.

[5] Trakhtenberg, I., Chronic Effects of Mercury on Organisms, Washington, D.C.: U.S. Dept. of Health, Education, and Welfare: Public Health Service: National Institute of Health: Superintendent of Documents, U.S. Government Printing Office, 1974.

Reading List

I. MERCURY TOXICITY

D'Itri, Frank M. and Patricia A. *Mercury Contamination (A Human Tragedy)* . John Wiley & Sons, Library of Congress, TD196.M38D57, 1977.

Fasciana, Guy S., *D.M.D. Are Your Dental Fillings Poisoning You?* Keats Publishing, Inc. (27 Pine St., P.O. Box 876, New Canaan, Conn. 06840), 1986.

Huggins, Hal, D.D.S., and Sharon, R.D.H., *It's All in Your Head.* P.O. Box 2589, Colorado Springs, CO 80901.

Montague, K., Montague, P. *Mercury.* Sierra Club (San Francisco, California), 1971.

Russell-Manning, Betsy *Candida (Silver Mercury Fillings and the Immune System).* Greensward Press (1600 Larkin St., Suite 104, San Francisco, CA 94109), 1985.

Russell-Manning, Betsy *How Safe are Silver (Mercury) Fillings?* Greensward Press (1600 Larkin St., Suite 104, San Francisco, CA 94109), 1985.

Stortebecker, Patrick, M.D., Ph.D. *Mercury Poisoning from Dental Amalgam — A Hazard to Human Brain.* Stortebecker Foundation for Research - s-18335 (Taby/Stockholm, Sweden), 1985.

Trakhtenberg, I.M. *Chronic Effects of Mercury on Organisms.* U.S. Dept. of Health Education and Welfare; Public Health Service; National Institutes of Health; Superintendent of Documents, U.S. Government Printing Office (Washington, D.C., 20402), 1974.

Tsubak, T. & Irukayama, K. *Minamata Disease.* Kodanska Ltd. and Elsevier Scientific Publishing Co. (Tokyo, Japan), 1977.

Ziff, Sam *The Toxic Time Bomb.* Aurora Press (205 3rd Ave., Suite 2A, New York, New York 10003), 1985.

II. NUTRITION AND GENERAL HEALTH

Airola, Paavo, N.D., Ph.D. *How to Get Well.* Health Plus Publishers (P.O. Box 22001, Phoenix, AZ 85028), 1974.

Jensen, Bernard, D.C. *My System.* (Bernard Jensen, D.C., Route 1, Box 52, Escondido, CA 92025), 1980.

Page, Melvin E., D.D.S. *Your Body is Your Best Doctor* Keats Publishing, Inc. (212 Elm St., New Canaan, Conn. 06840), 1972.

Price, Weston A., M.S., D.D.S., F.A.C.D. *Nutrition and Physical Degeneration.* The Price-Pottenger Nutrition Foundation, Inc. (P.O. Box 2614, La Mesa, CA 92041), 1945.

Tessler, Gordon S., Ph.D. *The Lazy Person's Guide to Better Nutrition.* Better Health Publishers (3925 W. Harmont Drive, Phoenix, Arizona 85051), 1984.

Index

Order Form

Name: _____

Address: _____
                          Street

_____
City   State   Zip

Please send me *The Complete Guide to Mercury Toxicity From Dental Fillings* by Joyal Taylor, D.D.S.

_____   copies @ $14.95 each .... $ _____

Shipping charges (see below) .... $ _____

6.5% Sales Tax to California addresses .... $ _____

TOTAL $ _____

Make checks payable to Scripps Publishing and send to:

> Scripps Publishing
> 9974 Scripps Ranch Blvd.
> San Diego, CA 92131

Shipping:

( ) **Book Rate:** $2.00 for the first book and $.75 for each additional book (surface shipping may take three to four weeks in the U.S.; up to four months foreign)

( ) **Air Mail:** $3.00 per book

Charge to my   VISA _____ MC _____ AE _____

Card #: _____

Expiration Date: _____

Signature: _____